SpringerBriefs in Anthropology

The SpringerBriefs in Anthropology series presents concise summaries of cutting-edge research and practical applications in all subfields of anthropology. Featuring compact volumes of 85 to 125 pages, the series covers a range of content from professional to academic. Typical topics might include: a snapshot of a hot or emerging topic, a contextual literature review, timely report of state-of-the art analytical techniques, in-depth case study, presentation of core concepts that students must understand in order to make independent contributions. Briefs allow authors to present their ideas and readers to absorb them with minimal time investment. Briefs will be published as part of Springer's eBook collection, with millions of users worldwide. In addition, Briefs will be available for individual print and electronic purchase. Briefs are characterized by fast, global electronic dissemination, standard publishing contracts, easy-to-use manuscript preparation and formatting guidelines, and expedited production schedules. We aim for publication 8–12 weeks after acceptance. Both solicited and unsolicited manuscripts are considered for publication in this series. Briefs can also arise from the scale up of a planned chapter. Instead of simply contributing to an edited volume, the author gets an authored book with the space necessary to provide more data, fundamentals and background on the subject, methodology, future outlook, etc.

Noor Ashikin Said

Cultural Translation in Technology Design

When Technology Travels

Springer

Noor Ashikin Said
MPU Department
Xiamen University Malaysia
Sepang, Selangor, Malaysia

ISSN 2195-0806 ISSN 2195-0814 (electronic)
SpringerBriefs in Anthropology
ISBN 978-3-032-13750-0 ISBN 978-3-032-13751-7 (eBook)
https://doi.org/10.1007/978-3-032-13751-7

This Springer imprint is published by the registered company Springer Nature Switzerland AG
The registered company address is: Gewerbestrasse 11, 6330 Cham, Switzerland

To Zaharul, Maya Adira, Zara Arissa and Lisa Zahira: the four points of my life's compass—though sometimes you spin it so much, I get dizzy.

Foreword

Much have been said and written on the relationship between culture and technology transfer, between developed and developing countries, between industries, from research institutions to commercial businesses. This book addresses all the three dimensions, using Malaysia as a focus of its empirical case study, with a catchy title *Cultural Translation in Technology Design—When Technology Travels* written by Dr. Noor Ashikin Said based on her doctoral research.

Malaysia is not new in the game of technology transfer heavily set in a cultural context famous for its multicultural and multiethnic diversity that shaped its labor force and organizational structure. For instance, language and dialects offer direct challenges in communicating the technology transfer process. Technology transfer as Malaysia had experienced it is not a straightforward process, as the narrative in this brilliant book reveals.

Malaysia has moved from the agricultural push to industrialization, albeit slowly, since the 1960s. The government has consistently used technology transfer as a central strategy to transition the nation into a high-income, developed economy. The big move toward industrialization was in the 1980s when Malaysia became involved in big industries, in the beginning mainly in automative industry, with the production a "national car" called Proton Saga, that uses Mitsubishi engine and the rest of the parts produced in Malaysia. It was a Malaysia-Japan joint industrialization effort set in a technology transfer and culture setting called Look East Policy (LEP).

In her detailed study, the first ever in Malaysia, she focuses on the aerospace sector in Malaysia in particular the transfer of Airbus technology to the production of composite parts of its airplanes using Virtual Reality (VR) training tool it developed in Europe. She concluded in her careful and in-depth study that "*while the VR system worked flawlessly in Europe, it faltered in Malaysia not due to technical flaws but because it carried hidden cultural scripts – expecting autonomy, linear learning and individual trial-and-error... assumptions clashed with local workshop practices grounded in hierarchy, tacit knowledge and collective apprenticeship...*"

Applying Alain Wisner's "anthropotechnology" framework in her study, she redefines technology transfer as negotiation not a delivery. As such, she emphasizes that "tools are not only technical artifacts but carriers of cultural imaginaries." The findings in the book *Cultural Translation in Technology Design—When Technology Travels* written by Dr. Noor Ashikin Said is relevant not only to Malaysia but also globally wherever and whenever technology transfer meets culture. An excellent contribution indeed to this field of study and should be of interest to both technology enthusiasts and social scientists.

Distinguished Professor
Dr. Shamsul Amri Baharuddin
FASc, Founding Director
Institute of Ethnic Studies (KITA)
Universiti Kebangsaan Malaysia
Bangi, Malaysia

Acknowledgments

This book represents more than an academic achievement; it is the story of resilience, sacrifice, and growth. What began as a doctoral thesis almost ten years ago has now found its form as a book, and along this long road, there have been countless challenges, unexpected turns, and generous hands that helped me rise each time I faltered.

When I embarked on my Ph.D., I never imagined that the journey would take nearly a decade. Balancing research and writing with the responsibilities of motherhood was not merely a logistical challenge—it was a test of spirit. There were days when I typed lines of my thesis with one hand while holding a baby in the other. Nights were often sleepless, not because of academic deadlines alone, but because of the cries of children needing comfort. In between preparing meals, school runs, and caring for my three daughters, I tried to carve out pockets of time to read, write, and think. The rhythm of my Ph.D. became inseparable from the rhythm of motherhood. My daughters grew up alongside this thesis; as they learned to walk, speak, and attend school, I was also learning to navigate the academic world, one page and one deadline at a time.

To every mother who has ever pursued higher education while raising children, I extend this acknowledgment to you as well. Your quiet strength, often unseen and uncelebrated, is the same strength that has carried me to this point.

I owe the deepest gratitude to my parents. My late mother, Rokiah Ramli, who is no longer here to see this moment, was the one who most wanted me to pursue education until the highest level, to become someone who could stand proudly as a lecturer and a scholar. Her prayers and dreams for me carried me through the hardest times, and I dedicate this work to her memory. My father, Said Hassan, who has worked humbly and tirelessly in the oil palm fields, has always asked about my studies, always reminding me of the value of patience and persistence. This achievement is as much his as it is mine.

This journey was made possible first and foremost through the guidance of my supervisor and mentor, Distinguished Prof. Datuk Dr. Shamsul Amri Baharuddin. His wisdom and generosity shaped not only this research but also my identity as a scholar. He did not only provide academic supervision; he offered patience when progress slowed, encouragement when I doubted myself, and above all, faith that this project was worth completing. I cannot count the number of times his words lifted me when I felt the burden was too heavy to bear. For this, I am eternally grateful.

I am also deeply indebted to my second supervisor, Dr. Eric Olmedo, who has been a steadfast mentor and guide throughout this process. His advice, insights, and constant encouragement were invaluable, especially during moments when I struggled to reconcile the demands of fieldwork, theory, and writing. His voice of reason and belief in my potential became an anchor throughout my academic journey.

My heartfelt appreciation goes to KITA, the Institute of Ethnic Studies, Universiti Kebangsaan Malaysia, which became my academic home during the years of my doctoral study. Serving as a research assistant in several projects not only provided me with valuable academic experience but also gave me a sense of belonging within a supportive and stimulating environment. I am grateful to my colleagues for their companionship, for the many conversations that sparked new insights, and for simply walking alongside me in this demanding path.

To the administrative staff at KITA, I owe special thanks. Research is never a solitary pursuit; behind every submission, every fieldwork clearance, and every meeting, there are people who work tirelessly to make things possible. Your quiet efficiency and kindness smoothed out countless obstacles. From arranging documents to providing logistical help, you made the daunting administrative side of research manageable. I carry immense gratitude for your work, which made all the difference in my ability to persevere.

To the engineers, managers, technicians and students in Malaysia's aerospace industry whom I had the privilege of studying and working with—this book belongs to you as much as it does to me. You welcomed me into your world, shared your experiences, and allowed me to witness how global technologies are localized through human ingenuity and cultural negotiation. Your patience, stories, and insights provided the very substance of this research. It is my hope that this book reflects, in some small way, the dedication and resilience you showed me in the field.

The writing of this thesis and book was never a linear process. It was fragmented by years of interruptions, personal responsibilities, and moments of doubt. There were times when I felt I might never complete it. But with each challenge, there were people who reminded me to keep going. I am profoundly thankful for the supportive community of scholars and friends who reminded me that academic work is not a solitary battle but a collective endeavor shaped by dialogue, collaboration, and kindness.

Beyond academia, this book is indebted to the unwavering support of my family. My husband, Zaharul Abdullah, has been my anchor in every sense of the word. He stood beside me through sleepless nights, moments of frustration, and times when I doubted whether I could finish. His patience, encouragement, and sacrifices made it possible for me to continue. Often, he carried the weight of responsibilities so that I could have the space to write. For this, I owe him more than words can express.

To my three daughters—Maya Adira, Zara Arissa, Lisa Zahira—who grew up in the shadow of this thesis, you are the quiet heroes of this story. You endured a mother who was often distracted, tired, and preoccupied. You gave me joy in the midst of exhaustion, laughter in moments of despair, and love that reminded me why perseverance mattered. Each of you, in your own way, gave me strength. This book is dedicated to you, not only because you lived through it with me, but because you are my greatest inspiration.

Completing this book has also been a process of personal growth. The journey from a struggling Ph.D. student to a lecturer has been long and difficult, but it has taught me resilience, humility, and the value of persistence. Today, as I stand in front of my own students, I carry with me the lessons of these years—the importance of patience, the courage to keep going, and the knowledge that intellectual work is inseparable from the personal lives we lead.

Finally, I would like to thank the anonymous reviewers and the editorial team at Springer. Their thoughtful feedback and careful reading pushed me to sharpen my arguments, clarify my prose, and rethink certain assumptions. The final version of this book is undoubtedly stronger because of their efforts. All errors that remain are, of course, my own responsibility.

In closing, I look back on this journey not simply as the completion of a thesis, but as the story of a life lived across different roles—mother, wife, daughter, researcher, assistant, teacher. This book holds within it not just arguments and theories, but also sleepless nights, moments of despair, and the triumph of persistence. To everyone who walked with me, carried me, or simply waited patiently for me to arrive at this point, I extend my deepest gratitude. This book is not mine alone.

Contents

List of Figures

List of Tables

Chapter 1
Introduction: Technology Is Never Universal—Theoretical Foundations

Abstract This chapter argues that technology is never universal but always situated in culture. Drawing on the Airbus case of Virtual Reality (VR) training tool transfer to its Malaysia supply chain, it challenges the persistent assumption that tools can be exported seamlessly across contexts. While the VR system worked flawlessly in Europe, it faltered in Malaysia not due to technical flaws but because it carried hidden cultural scripts—expecting autonomy, linear learning and individual trial-and-error. These assumptions clashed with local workshop practices grounded in hierarchy, tacit knowledge and collective apprenticeship. Building on anthropotechnology, a framework pioneered by Alain Wisner, the chapter redefines technology transfer as negotiation rather than delivery. Tools are not only technical artifacts but carriers of cultural imaginaries, requiring adaptation and translation to resonate with new users. Situating the Malaysian aerospace sector within broader debates on postcolonial modernity and socio-technical hybridity, the chapter proposes an ethic of transfer that treats culture as foundational, listening as engineering and translation as design. In doing so, it lays the theoretical foundation for the book's core concept—the anthropotechnological islet—as a landing space where global systems can pause, adapt and take root through co-creation with local actors.

Keywords Technology transfer · Anthropotechnology · Cultural adaptation · VR training · Global–local interface

Introduction

Every tool is born in a culture, and every transfer is a translation. Yet in the contemporary imagination of globalization, technology is often treated as universal—portable, self-contained and frictionless. This assumption is nowhere more visible than in international collaborations where complex systems are "exported" across borders and expected to perform as seamlessly abroad as they did at home.

But my fieldwork in Malaysia's aerospace sector suggested otherwise. In 2018, I walked into a training room at Composites Technology Research Malaysia (CTRM),

N. Ashikin Said, *Cultural Translation in Technology Design*,
SpringerBriefs in Anthropology, https://doi.org/10.1007/978-3-032-13751-7_1

where workers were testing a new VR training module for aircraft composite assembly. The training system, imported through the Airbus-led Virtual Immersive Reality System for Training in Aerospace Manufacturing (VIRISTAM) project, had been designed and perfected in Toulouse. Engineers spoke proudly of its precision, sleek interfaces and step-by-step tutorials.

On paper, it was flawless. Yet what I saw in that training room was hesitation. A young technician slipped on the VR headset but froze when the system prompted him to proceed without waiting for his peers. He chuckled nervously, removed the headset and turned to his colleague: *"Betul ke macam ni?" ("Is this really how it's done?").* Another worker asked to repeat the task on actual equipment before returning to the virtual module. Trainers, too, seemed uneasy. Some quietly reverted to their tried-and-tested paper manuals, guiding workers with a mixture of demonstration and story. Here lies the puzzle that animates this book: why do tools that work flawlessly in one setting stumble in another? Why do expensive systems falter, not because of technical breakdowns, but because of subtle mismatches in assumptions about users, learning and work? The answer, I argue, lies in culture—not as an obstacle but as a constitutive dimension of technology itself. VR did not "fail" at CTRM because workers were resistant or incompetent. It faltered because it carried hidden expectations: that learners would be autonomous, individualistic and comfortable with linear trial-and-error in isolation.

These assumptions clashed with Malaysian workshop practices, where learning is often collective, hierarchical and embodied. Rather than treating this mismatch as a failure, this book reframes it as a site of negotiation. Technology transfer is not delivery but dialogue. Tools do not simply travel; they arrive bearing scripts, values and imaginaries. To make sense in new contexts, they require adaptation, translation and sometimes reinvention. This is the essence of anthropotechnology, a field pioneered by Alain Wisner (1983, 1995), which insists that technology design and transfer must attend to cultural, cognitive and ergonomic diversity. This opening chapter lays the theoretical groundwork. It begins by dismantling the myth of universality, then situates technology as culture in motion. It places the Malaysian aerospace sector within this discussion, highlighting why the local context complicates "plug-and-play" logics. Finally, it introduces anthropotechnology as an ethic of design across difference, setting the stage for the empirical chapters that follow.

The Fallacy of Universality

The belief that technology is universal is deeply rooted in modernization theory and development discourse. W.W. Rostow's (1960) Stages of Economic Growth proposed that technological adoption was a linear driver of modernization: import machines, train people and progress will follow. In this worldview, culture is incidental—a backdrop that should not interfere with the rational march of progress.

This mindset persists today in many international collaborations. Systems are imagined as self-contained packages. Aerospace VR modules, medical devices and agricultural machines—all are designed in one context and exported to another with only minor "localization" tweaks such as translation of instructions. The assumption is that efficiency, precision and rational design transcend cultural difference. Yet anthropologists and Science and Technology Studies (STS) scholars have long critiqued this view. Bruno Latour (1992) argued that technologies are never "just objects" but socio-technical hybrids: bundles of hardware and human expectations.

Madeleine Akrich (1992) introduced the concept of scripts: the ways designers inscribe assumptions about users into artifacts.

A VR headset is not neutral; it scripts a user who is confident, self-motivated and willing to engage with trial-and-error in private. Lucy Suchman (2007) showed that even seemingly intuitive interfaces are culturally specific. Her ethnography at Xerox PARC revealed that what engineers coded as "natural" interactions were opaque to users unfamiliar with their logics. Alain Wisner (1983, 1995) took this critique further into ergonomics and design. He observed that machines built in European industrial contexts often failed when transplanted to Latin America or Africa—not because workers were deficient, but because the machines had been designed with different bodies, rhythms and work cultures in mind. For Wisner, universality was a dangerous illusion. He coined anthropotechnology as both critique and method: a way of designing and adapting technologies across cultural differences.

Our case at CTRM echoed these insights. The VR tool assumed universality, but its "script" did not fit. What engineers saw as hesitation was not incompetence but a different pedagogy: respect for hierarchy, reliance on peer confirmation and tacit knowledge embedded in gesture rather than explicit instruction. The transfer faltered not at the level of code, but at the level of culture.

Technology as Culture in Motion

To understand why universality fails, we must treat technology itself as culture in motion. Tools are not merely functional objects; they are crystallizations of cultural imaginaries. The VR module in our case was born in France, a context shaped by Enlightenment rationality, Cartesian logic and industrial pedagogy emphasizing autonomy and efficiency. Its design reflected these logics: step-by-step sequences, individual assessment and immediate feedback.

These were not arbitrary design choices but reflections of cultural assumptions about how people should learn and perform. When this system arrived in Malaysia, it collided with different cultural logics. Malaysian workshop learning is often collective. Apprentices learn by watching seniors, imitating rhythms and absorbing tacit cues. Silence is not resistance but respect; hesitation is not incompetence but deference. The VR module, designed for individual trial-and-error, stripped learning from this social fabric. Such mismatches are not unique to aerospace. In agriculture, European ploughs introduced to African contexts often failed because they assumed certain soil conditions and farming practices that did not exist locally (Biggs, 1990). In digital education, e-learning platforms designed for self-paced, individual study often falter in collectivist contexts where group discussion is central (Selwyn, 2014). Across sectors, technologies stumble when they mistake cultural specificity for universality.

In Malaysia, this disjuncture was heightened by the country's layered post-colonial history. British industrial models, Islamic educational traditions and local apprenticeship cultures intersect in complex ways. As anthropologist A.B. Shamsul (1996) argued, Malaysian modernity is not a linear adoption of Western forms but a hybrid assemblage. Any imported tool must navigate this hybridity. Thus, technology transfer is not simply about installing machines. It is about negotiating cultural logics. The VR system was not just a headset and software; it was a carrier of pedagogical assumptions. Its failure to "fit" was not a technical glitch but a cultural misalignment.

Malaysia's Socio-Technical Landscape

Understanding why this misfit mattered requires situating the case in Malaysia's broader socio-technical landscape. Malaysia has positioned itself as an aerospace hub since the 1990s. The National Aerospace Blueprint (1997, updated 2015) articulated ambitions to become a global supplier of aircraft components and maintenance services. CTRM, established in 1990, became a flagship of this vision, supplying composite parts to Airbus and Boeing. With government backing, it invested heavily in technology, facilities and workforce development.

Yet the workforce itself reflects Malaysia's diversity. Recruits come from varied educational backgrounds: some trained in technical colleges, others with minimal exposure to advanced technologies. Language diversity—Malay, English, Tamil, Mandarin—shapes communication on the factory floor. Training styles mix British colonial legacies (structured manuals, exams) with local practices of apprenticeship and observation.

This hybridity creates both opportunities and frictions. On the one hand, Malaysian workers demonstrate remarkable adaptability, moving fluidly between different logics. On the other hand, imported systems that assume homogeneity often falter. For example, when Japanese kaizen methods were introduced in Malaysian factories in the 1990s, they were reinterpreted through local norms of hierarchy and collectivism (Rasiah, 2002). Similarly, ICT adoption in schools revealed that students and teachers often relied on collective study groups rather than individualized modules (Lim, 2007).

In this context, the VR training module at CTRM was not just a technical innovation. It was an epistemological encounter. It brought assumptions about autonomy and linear sequencing into a context where learning was relational, tacit and multilingual. Silence in the training room carried meaning: it was not "failure to engage" but an act of respect. Workers' preference for confirming with peers was not inefficiency but a strategy of collaborative learning. To ignore these dynamics is to risk not only inefficiency but alienation. Technologies that clash with local logics often produce disengagement, quiet resistance or abandonment. By contrast, those that adapt—through dialogue, iteration and cultural listening—stand a greater chance of taking root.

Anthropotechnology as Framework and Ethic

This is where anthropotechnology enters. Alain Wisner (1983, 1995) argued that successful technology transfer requires more than technical calibration; it requires cultural and cognitive adaptation. Anthropotechnology is both critique and method: it exposes the failures of universalist design and provides tools for adaptation across contexts.

The anthropotechnological approach resonates with participatory design (Schuler & Namioka, 1993), cultural ergonomics (Imada, 2001) and inclusive technology (Irani et al., 2010). All emphasize the situatedness of users and the necessity of co-design. But anthropotechnology brings a unique focus on global transfers and postcolonial contexts. It asks: how do tools designed in one world make sense in another? In practice, anthropotechnology operates through ethnography, iteration and dialogue. In the VIRISTAM project, this meant months of fieldwork in CTRM workshops, observing how workers engaged with VR, listening to their stories and capturing their tacit knowledge.

Based on the above discussion, an important question arises: how can these concepts be correlated and operationalized as analytical tools in this study? In response, the research is structured around five guiding pillars: (i) identifying the local social actors in Malaysia that act as intermediary agents of technology transfer; (ii) analyzing and understanding the sociological profile of the drilling technicians at CTRM; (iii) assessing how the use of the virtual reality training tool impacted the ergonomics of drilling technicians; (iv) examining the cognitive ability levels of those technicians; and (v) evaluating the communication codes that shaped collaboration on the factory floor. These objectives are bound together through the framework of anthropotechnology, which holds them in a single conceptual and ethical frame. Each objective is analyzed through specific indicators, detailed in Fig. 1.1.

At the micro level, the analysis focuses on the everyday lives of technicians: their sociological profiles, their cognitive rhythms and the communication codes that circulate on the shop floor. To trace the sociological profile, indicators such as background, training experience, ergonomic challenges and acculturation strategies are considered. Cognitive ability is explored through computerized tests that distinguish allocentric from egocentric orientations, shedding light on how individuals adapt to demanding precision work.

Communication codes—the tacit and explicit ways workers instruct, correct and coordinate—are equally central, for they reveal how technology is made meaningful in practice. At the meso level, attention shifts to the VR training tool itself, where impact is measured in terms of training time, cost reduction and safety outcomes. But this level also captures the wider institutional collaborations that mediate the transfer: AIRBUS, AMIC, CTRM, UM and UKM. Indicators here include the permeability

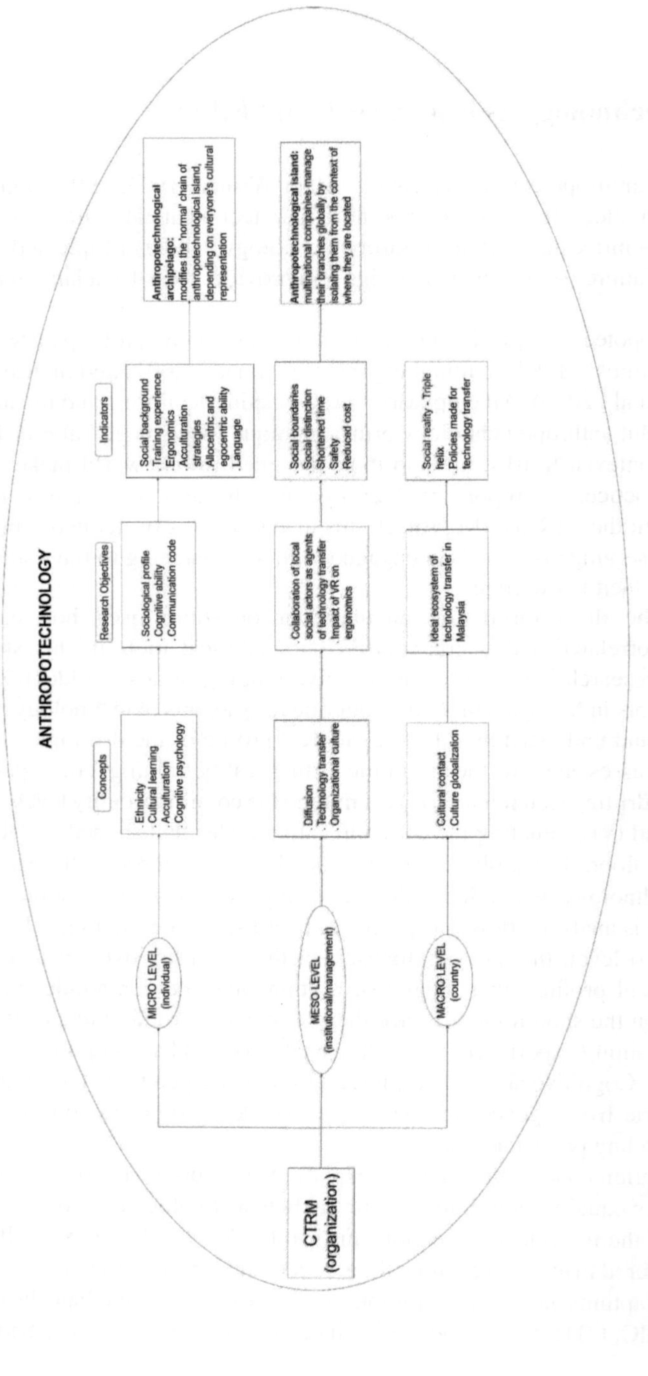

Fig. 1.1 Conceptual framework of cross-cultural technology transfer at CTRM (Author's own 2024)

of social boundaries, the persistence of distinctions and the emergence of stigma in cross-organizational exchanges.

At the macro level, the analysis situates these interactions within broader cultural and national frames. Indicators such as cultural globalization and "pictured social reality" are used to interpret how transcultural transfers ripple outward, reshaping not only organizations but also the narratives of national development. This study therefore rests on the claim that technology is never neutral but culturally embedded. Cultural contact alters the host country's social structure at multiple scales. To operationalize this, the theoretical framework maps the transfer process from the spark of innovation to the end-user, as shown in Fig. 1.2.

This model makes visible the architecture that guided the study. It traces the technology itself—VIRISTAM—from its origins in France, through the hands of AMIC as intermediary, into CTRM as end-user and across the multiple institutional nodes that shaped it (industry, government, university). It also recognizes that the transfer was never purely technical: AIRBUS sent experts, trained engineers and joined steering committees, while local partners translated, negotiated and reshaped the tool. The process was at once technical and cultural, and anthropotechnology provided the lens through which these entanglements could be observed.

By keeping CTRM involved throughout the design process, by modifying structures and flows to match cultural rhythms and by embedding the end-user's voice, the project illustrates Wisner's argument that adaptation is as important as invention. The concept of an anthropotechnological islet—land of adaptation where culture and technology meet—became a useful metaphor. This "islet" was the space where technicians' knowledge, engineers' expertise and institutional agendas met, producing a personalized technology rather than a transplanted one.

Toward a New Ethic of Transfer

The lesson of CTRM is simple yet profound: technology does not transfer by itself; people transfer it. The VR system faltered not because Malaysians were resistant, but because its design assumed a universal user who did not exist. Once workers' tacit practices were recognized, once silence was reinterpreted as deference rather than resistance, once trainers' insights were integrated, the system began to make sense. This reframing demands a new ethic of transfer. One that treats culture as foundational, not incidental; sees listening as a mode of engineering; understands translation as design and values negotiation over standardization.

Such an ethic does not romanticize cultural difference. It does not suggest that every tool must be reinvented from scratch. Rather, it insists that all tools are partial, situated and in need of dialogue. The CTRM case is not an exception but an exemplar. Around the world, technologies stumble when they ignore the in-between spaces where systems meet situations. The challenge is not to "fix" users but to design smarter—to co-create tools that resonate with local logics while retaining technical strength.

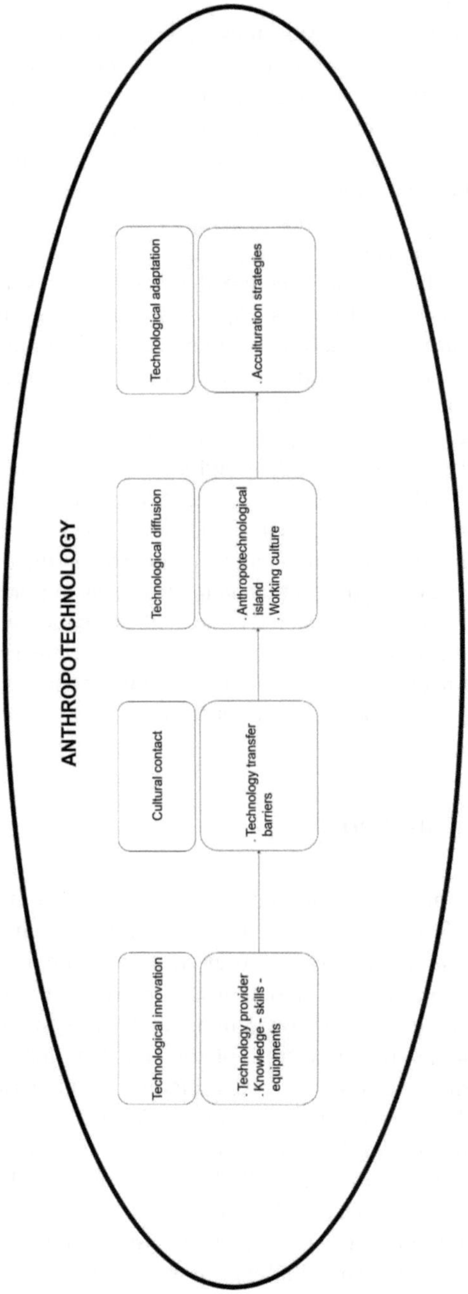

Fig. 1.2 Model of cross-cultural technology transfer at CTRM (Author's own 2024)

As this book unfolds, I will return to the CTRM floor, where VR met Malaysian rhythms, and where universality gave way to negotiation. The story begins not with seamless adoption, but with hesitation. And it is in those hesitations—those pauses, chuckles and whispered confirmations—that the real work of technology transfer takes place.

References

Akrich, M. (1992). The de-scription of technical objects. In W. Bijker & J. Law (Eds.), *Shaping technology/building society* (pp. 205–224). MIT Press.

Biggs, S. (1990). A multiple source of innovation model of agricultural research and technology promotion. *World Development, 18*(11), 1481–1499.

Dourish, P., & Bell, G. (2011). *Divining a digital future: Mess and mythology in ubiquitous computing*. MIT Press.

Imada, A. S. (2001). The rationale and tools of cross-cultural ergonomics. *Theoretical Issues in Ergonomics Science, 2*(3), 278–292.

Irani, L., Vertesi, J., Dourish, P., Philip, K., & Grinter, R. (2010). *Postcolonial computing: A lens on design and development*. In CHI '10 Proceedings.

Latour, B. (1992). Where are the missing masses? The sociology of a few mundane artifacts. In W. Bijker & J. Law (Eds.), *Shaping technology/building society*. MIT Press.

Lim, C. P. (2007). Effective integration of ICT in Singapore schools: Pedagogical and policy implications. *Educational Technology Research and Development, 55*(1), 83–116.

Rasiah, R. (2002). Systemic coordination and human capital development: Knowledge flows in Malaysia's MNC-driven electronics cluster. *Transnational Corporations, 11*(3), 89–130.

Rostow, W. W. (1960). *The stages of economic growth: A non-communist manifesto*. Cambridge University Press.

Said, N. A. (2024). *A cross cultural technology transfer in Malaysia aerospace industry: An anthropotechnological approach* [Unpublished doctoral dissertation]. Universiti Kebangsaan Malaysia.

Schuler, D., & Namioka, A. (1993). *Participatory design: Principles and practices*. CRC Press.

Selwyn, N. (2014). *Distrusting educational technology: Critical questions for changing times*. Routledge.

Shamsul, A. B. (1996). Nations-of-intent in Malaysia. In S. Tonnesson & H. Antlöv (Eds.), *Asian forms of the nation* (pp. 323–347). Curzon.

Suchman, L. (2007). *Human-machine reconfigurations: Plans and situated actions* (2nd ed.). Cambridge University Press.

Wisner, A. (1983). *Ergonomie et anthropotechnologie*. CNRS.

Wisner, A. (1995). *Understanding problem building: Ergonomics work analysis*. Taylor & Francis.

Chapter 2
When Technology Travels: A Case in Translation

Abstract This chapter explores the dynamics of technology transfer between Airbus and CTRM through the introduction of a VR training tool. It argues that technology does not simply move intact across borders but must be reinterpreted within local contexts of practice. While Airbus engineers viewed the VR simulator as a rational solution to skill gaps, CTRM technicians approached it through tacit routines, embodied knowledge and cultural values of hierarchy and respect. The chapter situates these encounters within the broader literature on technology transfer and sociotechnical systems, showing how misalignments in training, communication and organizational culture reshaped the tool's use. Drawing on ethnographic observations, it highlights the role of cultural mediation in bridging assumptions across national and institutional boundaries. The analysis reframes technology transfer as a process of cultural translation, where success depends not only on technical design but also on shared meanings negotiated on the factory floor.

Keywords Technology transfer · Virtual reality · Cultural translation · Aerospace industry · Ethnography

Introduction

The promise of technology transfer is often narrated as a story of progress: a tool developed in one context, refined by experts and then passed on to another where it is expected to deliver similar gains. The assumption embedded in this narrative is that technologies, like blueprints, travel neatly across borders. But the case of CTRM in Melaka complicates this picture. What unfolded when Airbus attempted to introduce a VR training tool in Malaysia was not a failure of engineering. It was, more subtly, a problem of translation. Not of words, but of rhythms, gestures, expectations and identities.

In 2015, CTRM was at a critical juncture. The facility had grown into one of Malaysia's most visible success stories in aerospace manufacturing, producing

N. Ashikin Said, *Cultural Translation in Technology Design*,
SpringerBriefs in Anthropology, https://doi.org/10.1007/978-3-032-13751-7_2

composite structures for the Airbus A320 family and positioning itself as a dependable partner in the global supply chain. The factory floor gleamed with state-of-the-art equipment, and the carbon-fiber components emerging from it met exacting international standards. Yet behind this surface of precision lay a quieter challenge. Skilled technicians, trained at significant cost, were leaving at higher-than-expected rates. Turnover was eroding continuity, inflating recruitment budgets and straining morale among trainers and supervisors.

Each resignation initiated a cycle that became familiar: an exit interview, hurried hiring, onboarding of replacements and the careful retraining of new recruits. Just as a team regained stability, another departure would unsettle the balance. The problem was not the quality of training—Malaysia had invested heavily in human capital since the 1997 National Aerospace Blueprint—but the retention of embodied know-how. Tacit skills are fragile; they cannot be archived like documents or codified like procedures (Nonaka & Takeuchi, 1995). They live in hands, in eyes, in unspoken timing between a senior technician's gesture and a junior's imitation. When those hands walked out of the factory, the knowledge left with them.

It was in this context that Airbus proposed an intervention: a VR training tool originally designed in Toulouse. In France, the tool had shown remarkable promise. With a headset and controllers, trainees could simulate assembly tasks, practice sequences and receive immediate feedback without consuming physical resources or disrupting the production floor. The appeal to CTRM managers was obvious. If successful, the tool could standardize training, accelerate onboarding and reduce the reliance on overstretched human trainers. It was, in managerial language, a scalable solution to a pressing human problem.

Yet the introduction of the VR module did not unfold as expected. Technicians were not hostile to technology; many were avid smartphone users, comfortable with digital interfaces and familiar with gaming consoles. But something about the VR system felt alien. Some removed the headset halfway through a session, complaining of dizziness or disorientation. Others hesitated before performing tasks, their movements slower and less assured than in real life. A few disliked being alone in the simulation, finding the solitary environment unsettling compared to the collaborative atmosphere of the training floor.

These reactions were puzzling to the engineers. The software functioned flawlessly. The sequences were accurate. The error-detection algorithms were precise. Why, then, did technicians struggle? To answer this, we must look not at the hardware but at the assumptions embedded in its design.

The Material Architecture of VIRISTAM

The hardware was more than a collection of wires, trackers and headsets. It was the scaffolding through which technicians encountered the virtual, and thus it shaped how learning itself was experienced. Realistic Human Experiment Analysis (RHEA), the platform for VIRISTAM, relied on three interdependent systems—tracking, interaction and visualization—that together made immersion possible. To appreciate the responses of technicians later described, we must first understand this architecture.

What follows is a closer look at the hardware environment of RHEA, the platform on which VIRISTAM was built. At its core, the VIRISTAM system relied on a carefully assembled hardware ecosystem. Like any immersive environment, the experience was only as convincing as the physical instruments that mediated between human perception and the virtual model. The hardware required to run RHEA consisted of three key subsystems: tracking, interaction and visualization. Together, they created the interface through which technicians were expected to train, adapt and ultimately embody Airbus drilling protocols.

Figure 2.1 shows the basic configuration of the VR setup when projected on a power wall. In other cases, the same system could be run through a head-mounted display (HMD), replacing screen and projector with a device that enveloped the user's field of vision. Each configuration carried different implications for how technicians encountered the tool: on the power wall, the VR remained a shared and semi-public experience; on the HMD, it became intimate, immersive and isolating.

To sustain immersion, the system had to know precisely where the user was and how they moved. For the VIRISTAM project, this was accomplished through ARTTRACK technology: six Arttrack2 cameras, a central controller and DTrack2 software that registered every tracker in the environment as pictured in Fig. 2.2.

The system came with sixteen trackers, though nine were deployed in Melaka—positioned on head, hands, waist, feet, flystick controller, the Haption haptic arm and the HMD. Each tracker translated physical motion into digital data, mapping the technician's body into the simulation. The logic was simple: by knowing where the body was, the system could build a credible illusion of presence. But the practice

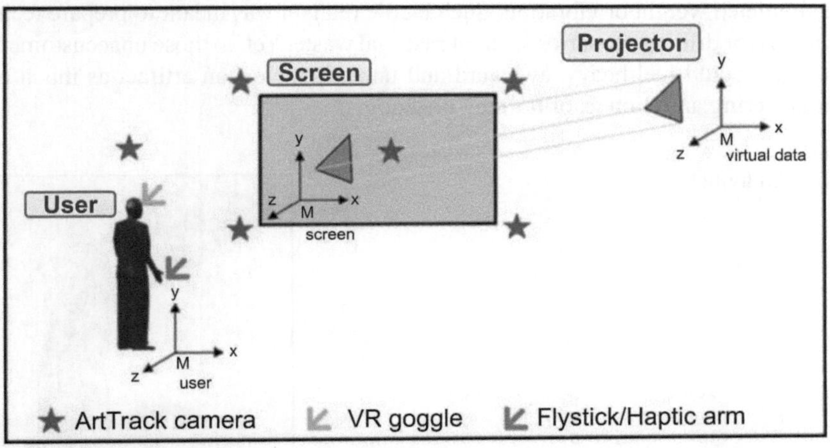

Fig. 2.1 Hardware setup

Fig. 2.2 VR goggles with tracker

was less straightforward. The weight of trackers, the discipline of standing within calibrated zones and the translation lag between movement and feedback all subtly shaped the technicians' experiences. The promise of VR is not just to see but to act. For this, VIRISTAM relied on a set of interactive devices:

a. ART Flystick (navigation and manipulation system)

   The flystick was designed as a universal controller (Fig. 2.3). It allowed users to fly or walk through the simulation, grab virtual objects, measure dimensions and even construct simple shapes. In the engineers' imagination, this tool empowered technicians to explore freely. In practice, however, its layered functions often overwhelmed those who were still unfamiliar with navigating three-dimensional virtual space.

b. Haptic Arm (Haption force-feedback system)

   If the flystick mediated vision and movement, the haptic arm mediated touch. With its force-feedback capacity, users could "feel" resistance from a virtual object (Fig. 2.4). Pick up a digital component, and the arm would push back with a simulated weight or vibration. Such tactile realism was meant to prepare technicians for drilling scenarios without material waste. Yet, to those unaccustomed, the arm could feel heavy, awkward and unpredictable—an artifact as much of engineering ambition as of training pedagogy.

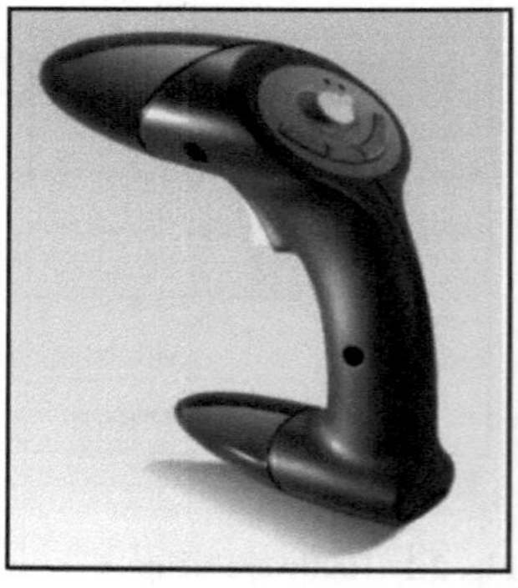

Fig. 2.3 Art flystick

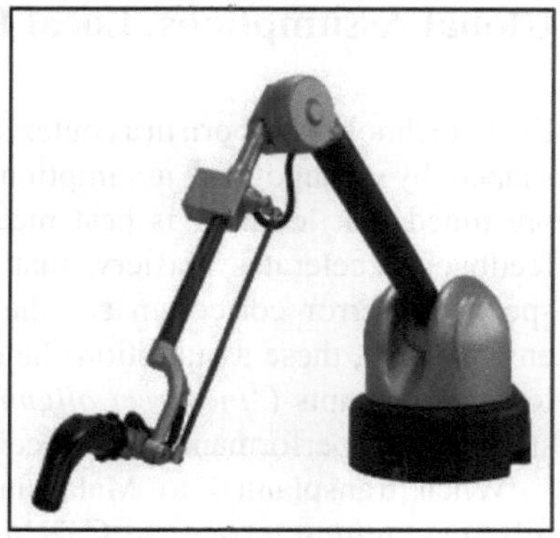

Fig. 2.4 Haptic arm

c, Head-mounted display (HMD)

The HMD offered the deepest immersion (Fig. 2.5). Unlike the power wall, which invited group observation, the headset enclosed the user in a 360-degree stereoscopic world. Whatever the user saw was simultaneously mirrored on an external monitor for others to follow, though only from the user's vantage point. This created both intimacy and disconnection: the technician was "inside," while colleagues remained spectators on the outside. For some, this was exhilarating; for others, disorienting.

Yet hardware alone did not determine outcomes. However sophisticated, the cameras, sensors and haptic devices merely provided the stage. What unfolded on that stage depended on the assumptions encoded into the software and more deeply, on the cultural logics of learning that framed how trainees engaged with it. This brings us to the question of pedagogy: the global assumptions built into the Airbus tool, and the local realities it encountered in Melaka.

Fig. 2.5 Head mounted display

Global Assumptions, Local Realities

Every technology is born in a context. The Airbus VR tool was developed in Toulouse, shaped by pedagogical assumptions common in European technical training. It presumed that learning is best measured individually, that direct and immediate feedback accelerates mastery, that abstract interfaces pose no obstacle and that speed and error correction are the primary indicators of progress. In its native environment, these assumptions held true. Trainees in France responded positively to blunt prompts ("*Incorrect alignment—try again*"), and the system successfully standardized performance across cohorts (Bailenson, 2018).

When transplanted to Malaysia, however, these assumptions encountered a different cultural terrain. At CTRM, training was not only about mastering procedures but also about navigating relationships. New recruits learned less from manuals than from watching seniors, imitating movements and rehearsing tasks collectively. Silence was not a void but a signal of respect; questions were asked indirectly, often during tea breaks or through peers rather than directly to superiors. Feedback was rarely confrontational. A trainer might gently say, "*Cuba perhatikan sekali lagi*," using tone and gesture rather than explicit correction.

This rhythm of learning was slower but socially attuned. It allowed trainees to build confidence without fear of embarrassment, preserved hierarchies without humiliation and fostered solidarity among peers. What looked to an outsider like inefficiency was, in fact, a finely calibrated ecology of learning. The VR tool, with its solitary interface, abstract commands and relentless pace, disrupted this ecology. Interviews revealed deeper insights.

One technician explained, "*I don't like training alone. In the factory, I always have a buddy to check me. Here, in VR, I feel like I might make a mistake and no one is there to guide me.*" Another said, "*I cannot feel with my hands. In real life, I know if the layer is right because of touch. Here, I just guess.*" These were not signs of incompetence but of misalignment. The tool carried the pedagogical DNA of Toulouse and had not been re-coded for Melaka. The problem was not transfer but translation. Understanding translation, however, required more than technical fixes. It demanded a closer look at everyday practices of learning, the small gestures and rhythms that engineers often overlooked.

The Slow Work of Translation

Recognizing this gap required slowing down. Translation here was not immediate; it unfolded through the slow work of observing, listening and reshaping the tool in dialogue with its users. Each adjustment, while minor in coding terms, represented a significant cultural translation of pedagogy. The adaptation process shifted from engineering to ethnography. Project teams began observing training sessions, noting how silence structured interaction, how peers exchanged glances before repeating a task, how stories and anecdotes carried procedural knowledge across shifts. Trainers were not just instructors; they were narrators, embedding technical steps in cultural forms of recall.

To see how this played out in practice, it helps to look more closely at the training modules themselves. Two use cases structured the VR program. The first reproduced the flow of carbon-fiber composite production, where trainees pushed trolleys—nicknamed "Christmas trees"—from the before-cure zone into the autoclave, and then onwards to the after-cure zone. Instruction boards lined the pathway, guiding each stage of the process with text and voice prompts. The logic was procedural and linear, designed to test whether technicians could follow sequential commands and complete the workflow without deviation. The second use case shifted focus from process to skill. Here, trainees drilled carbon-fiber panels clamped to a workbench, practicing alignment, rotation speed and pressure.

Precision and ergonomics were central, as were lessons in safety, tool wear and posture. In both scenarios, the VR system translated embodied shop-floor practices into discrete virtual steps. What counted as "success" was not the subtle give of material beneath one's hands, but the proper execution of coded instructions. For many technicians, this was their first encounter with such an environment and their reactions revealed how differently learning was calibrated on the factory floor versus in the simulation.

Workshops with technicians revealed preferences often invisible to managers. Prompts were rewritten in relational language. Voiceovers were recorded in local accents, with tones closer to encouragement than correction. Interface designs allowed for pauses, repetition and even peer input. Motion capture sequences simulated not only the task but also its rhythm, showing how long a senior might linger before moving to the next step. These adjustments may appear cosmetic, but they were cognitive. They shifted the system's logic from an individualistic, task-oriented pedagogy to one aligned with collective, relational and reflective learning practices.

The result was a tool that felt less foreign. Trainees began to complete modules with fewer interruptions. Trainers reported fewer follow-up questions. Workers started describing the system not as a machine but as a mentor. The lesson here is broader. Technologies do not travel empty; they carry cultural assumptions, ergonomic logics and pedagogical rhythms. When they arrive in new settings, they encounter equally robust local practices. The meeting of these logics is not friction to be eliminated but intelligence to be harnessed. What looks like failure may be a clue to how people actually learn.

Culture as Intelligence

The CTRM case suggests a reframing. Rather than treating culture as interference, we can treat it as intelligence. Silence, storytelling and group-based repetition—often dismissed by efficiency-driven frameworks—are not obstacles but resources. They are strategies through which knowledge is preserved, trust is built and learning is sustained. To design adaptive technologies, engineers must listen not only to what users say but to how they move, pause and relate.

This insight is central to what I call an anthropotechnological islet: a landing zone where global tools learn to fit local hands. It is not a wholesale rejection of imported systems but a recognition that adaptation requires humility. The work of translation is slow, iterative and dialogic. It involves not only reprogramming software but also rethinking what counts as learning, what counts as efficiency and what counts as success. As the adapted VR tool re-entered CTRM, it did so not as an imposition but as a partner. It became part of a hybrid pedagogy, combining global precision with local rhythms. The technology had not simply been transferred; it had been translated.

Conclusion

This chapter has traced the journey of a VR training tool from Toulouse to Melaka, from promise to puzzle to adaptation. It has shown that technology transfer is never a simple handover. It is a cultural encounter where assumptions collide, where silence speaks and where translation—not transfer—determines success.

The lesson extends beyond aerospace. Every global tool, whether a machine, a platform or a protocol, carries the cultural DNA of its origin. To succeed elsewhere, it must be unpacked, re-narrated and re-aligned. In this sense, technology is less like a universal language and more like a dialect—always intelligible, but only after careful translation. For CTRM, the VR project became more than a training initiative. It was a reminder that culture is not noise in the system but the medium through which systems become meaningful.

For scholars of technology transfer, it offers a methodological cue: to attend not only to contracts and metrics but to pauses, gestures and stories. For designers, it is a challenge: to build with cultural intelligence, not despite it. As technology travels, they bring baggage. Unless that baggage is unpacked, the tool may never fully arrive.

References

Bailenson, J. (2018). *Experience on demand: What virtual reality is, how it works, and what it can do*. W. W. Norton.

Gibson, C., & Zellmer-Bruhn, M. (2001). Metaphors and meaning: An intercultural analysis of the concept of "teamwork." *Administrative Science Quarterly, 46*(2), 274–303.

Hofstede, G. (2010). *Cultures and organizations: Software of the mind* (3rd ed.). McGraw-Hill.

Khairuddin, N., Mohd Yusof, R., & Ahmad, Z. (2015). Malaysia's aerospace strategy: Building a regional hub. *Journal of Aerospace Policy, 7*(2), 45–61.

Lave, J., & Wenger, E. (1991). *Situated learning: Legitimate peripheral participation*. Cambridge University Press.

Lim, S. (2017). Learning in Malaysian technical training: Practices and norms. *Asian Journal of Education and Work, 4*(1), 23–39.

Nonaka, I., & Takeuchi, H. (1995). *The knowledge-creating company*. Oxford University Press.

Orr, J. E. (1996). *Talking about machines: An ethnography of a modern job*. Cornell University Press.

Said, N. A. (2024). *A cross cultural technology transfer in Malaysia aerospace industry: An anthropotechnological approach* [Unpublished doctoral dissertation]. Universiti Kebangsaan Malaysia.

Suchman, L. (2007). *Human–machine reconfigurations: Plans and situated actions*. Cambridge University Press.

Chapter 3
Anthropotechnology: Reframing Technology Transfer as a Human Encounter

Abstract This chapter explores anthropotechnology as a framework for understanding technology transfer, with a focus on the Airbus–CTRM collaboration in Malaysia. It argues that technology transfer is never a neutral, purely technical process, but always mediated by cultural assumptions, practices and negotiations. Drawing on the insights of Alain Wisner and subsequent developments in ergonomics and anthropology, the chapter highlights how cultural brokers facilitated the adaptation of European aerospace technologies into Malaysian contexts. Through the case study of VR training and composite manufacturing, it illustrates the tensions and creative adaptations that arise when imported systems meet local practices. The chapter concludes by framing technology transfer as a cultural encounter, in which hybridity rather than replication becomes the marker of success.

Keywords Anthropotechnology · Technology transfer · Cultural brokers · Aerospace industry · Malaysia

Introduction

Technology transfer is often imagined as a seamless relay: a tool perfected in one place is simply handed over to another, and productivity follows. But reality unsettles this neat picture. Why do technologies that function flawlessly in Toulouse stumble in Melaka? Why does a training tool, designed with precision and tested to industrial standards, suddenly feel alien, unusable or even resisted when it crosses borders? The Airbus–CTRM collaboration offers a compelling answer: the obstacle does not lie in the machine itself but in the human and cultural environment into which it lands.

This chapter reframes technology transfer through the lens of anthropotechnology—a field that refuses to separate machines from the people who use them. Anthropotechnology does not treat technology as a neutral object, nor does it see workers as passive recipients. Instead, it situates technical systems within living cultures, workplaces and social imaginaries (Wisner, 1997). In doing so, it reveals a

N. Ashikin Said, *Cultural Translation in Technology Design*,
SpringerBriefs in Anthropology, https://doi.org/10.1007/978-3-032-13751-7_3

paradox: the success of cutting-edge tools depends less on their technical design than on the ability to mediate between worlds—between the global logics of aerospace engineering and the local rhythms of Malaysian work culture.

The story that unfolds in this chapter begins with the theoretical roots of anthropotechnology, born out of French ergonomics and the study of "man–machine systems" in the mid-twentieth century. But theory alone cannot carry us far. To understand what anthropotechnology looks like in practice, we turn to the Airbus–CTRM case, where engineers, trainers and workers navigated not only the complexities of composite materials and VR training modules but also the subtler frictions of language, hierarchy and habit. In this space, cultural brokers—those who straddle both technical and cultural domains—emerged as the decisive actors.

Anthropotechnology, as this chapter will argue, offers more than a vocabulary; it provides a methodological stance. It teaches us to look not at the machine in isolation, nor at the worker in abstraction, but at the encounter where both meet. And in that encounter, culture is never a background variable—it is the medium through which technology is translated, adapted and sometimes transformed.

From Ergonomics to Anthropotechnology

The roots of anthropotechnology lie in the French tradition of ergonomics, particularly in the work of Alain Wisner, who coined the term in the 1970s. Ergonomics had long sought to optimize the fit between worker and machine, but in industrial contexts dominated by Western assumptions, "fit" was often defined in universal terms. Anthropotechnology marked a departure. It recognized that the "human" in human factors was not a generic body but a situated being shaped by culture, social structure and history (Wisner, 1997).

In its early formulations, anthropotechnology emerged from postcolonial encounters. French engineers, tasked with implementing industrial projects in Africa, Latin America and Asia, realized that machines designed for European conditions often failed abroad—not because of flaws in engineering, but because of mismatches in work organization, climate or cultural practices. A drill that worked in France overheated in tropical humidity. A factory schedule optimized for Parisian routines clashed with local rhythms of prayer, family or agriculture (Rogalski, 1991). These were not trivial obstacles but deep reminders that technology is never "just technical."

This realization gave birth to a new imperative: technologies must be re-engineered through culture, not in spite of it. Anthropotechnology, therefore, is less about transplanting tools than about translating systems—aligning technical structures with the lived realities of workers. In this sense, anthropotechnology prefigured contemporary debates in science and technology studies (STS), which likewise emphasize the co-construction of technology and society (Akrich, 1992; Latour, 1987).

Yet anthropotechnology retains a distinctive flavor. Where STS often hovers at the level of discourse and networks, anthropotechnology insists on the concreteness of work—the gestures of a hand, the bodily strain of lifting and the micro-rhythms of training sessions. This is what makes it particularly relevant to the Airbus–CTRM case, where the gap was not only cultural in the abstract but embodied in how workers learned, moved and interacted with machines.

The anthropotechnological method rests on a principle of co-construction: technology and society are never fully separable but rather shape and reshape one another. To understand this co-construction, we need to look beyond mere ergonomic adaptation to include institutional and cultural adaptation. The VR training tool in the Airbus–CTRM case is instructive here. It was designed in Toulouse for a European audience of engineers accustomed to direct, individualized learning and embedded within a broader culture of standardized certification. When the same tool landed in Melaka, the assumption that knowledge is transmitted one-to-one was disrupted. Local trainees came from educational traditions emphasizing collective problem-solving and hierarchical guidance. Trainers often preferred to work in groups, and supervisors were expected to mediate between tool and trainee. The anthropotechnologist's eye sees these moments not as failures of technology but as clues about divergent epistemologies of learning (Wisner, 1997).

Cultural Brokers and Mediation

At this juncture, the role of cultural brokers becomes central. In the literature on migration and cross-cultural management, brokers are those who translate meanings, bridge institutions and mediate between otherwise incompatible systems (Bochner, 2016; Steyaert & Janssens, 1997). In anthropotechnology, brokers play an analogous role: they are the engineers, middle managers or even shop-floor operators who interpret the imported system and adjust it for local resonance.

At AMIC, the broker was senior engineer who had trained in Europe but was deeply embedded in the local work culture. They understood both the Airbus certification regime and the Malay workplace ethos of respect, indirect communication and collective responsibility. Rather than simply enforcing the VR module as written, they re-contextualized it: adding collective debriefings, creating informal peer-teaching circles and softening the disciplinary edge of the European system with culturally resonant metaphors. In doing so, they preserved technical fidelity while creating social legitimacy.

This example reveals an important methodological lesson. Anthropotechnology is not just about identifying "mismatches" and correcting them; it is about recognizing that all technologies carry cultural inscriptions. A VR headset is not a neutral device—it embodies assumptions about attention span, comfort with virtuality and even tolerance for isolation. Here, anthropotechnology provides a diagnostic lens: the problem is not that Malaysian workers were "resistant" to VR, but that VR as designed reflected only one cultural model of learning.

Technology Transfer as Translation

The deeper implication is that technology transfer is always technology translation. The Airbus–CTRM case mirrors older anthropotechnological interventions in Africa and Latin America, where imported factory layouts or safety protocols had to be reworked to align with local rhythms, taboos and communal practices (Wisner, 1983, 1994). The key difference is that in aerospace, the tolerance for deviation is far smaller. Planes cannot be "almost safe." Thus, brokers at CTRM operated within a tight corridor: they had to adapt practices to local culture while maintaining uncompromised compliance with Airbus's airworthiness standards. This tension—between cultural adaptability and technical rigidity—is at the heart of anthropotechnology in high-stakes industries.

Consider, for example, the issue of language. Airbus delivered documentation and VR scripts in English, assuming global comprehensibility. In practice, workers toggled constantly between English technical jargon, Bahasa Malaysia and sometimes Mandarin. Direct translation often produced confusion: a single Airbus term might have no equivalent in Malay, or worse, a misleading one. Cultural brokers stepped in to produce living glossaries, where meaning was negotiated. Rather than a one-off translation, this was a continuous process of linguistic adaptation, shaped by trust and reciprocity. From an anthropotechnological perspective, language here is not just a medium of information but a technology of coordination—itself subject to design, breakdown and repair.

The methodological point is clear: anthropotechnology requires attention to the lived micro-practices of adaptation. It is not sufficient to map institutional structures; one must observe the brokers in action, the subtle adjustments and the reframing of rules. In Melaka, a broker might explain a complex Airbus procedure not by literal translation but through analogy to local practices of craft or agriculture. Such strategies embody what Wisner called "the anthropological dimension of ergonomics" (Wisner, 1988): the recognition that tools only work when they resonate with embodied cultural logics.

The Airbus–CTRM collaboration provides a striking illustration of this dilemma. When Airbus introduced advanced composite technologies to CTRM in Malaysia, the technical protocols were delivered with precision. Manuals were translated, training sessions were scheduled and engineers were dispatched to demonstrate best practices. Yet the outcomes were uneven. Some processes were absorbed seamlessly, while others stalled or produced errors that were difficult to diagnose. On paper, the transfer had been executed flawlessly. In practice, the technology behaved as if it were resistant to transplantation.

This resistance cannot be reduced to incompetence or lack of resources on the receiving side. Instead, it reflects the collision between two worlds of meaning: one shaped by European industrial culture, with its assumptions of standardization, punctuality and linear problem-solving; the other informed by Malaysian workplace dynamics, where hierarchy, interpersonal obligations, and improvisation often take precedence. Anthropotechnology suggests that technologies cannot be insulated from these cultural contexts. They do not float above society; they sink into it, shaped by the expectations, fears and values of the humans who operate them (Béguin & Weill-Fassina, 1997).

One of the most important contributions of anthropotechnology is its recognition that adaptation flows both ways. It is not only that local workers must adapt to imported technologies; technologies themselves must be reshaped to fit the contours of local cultures. In Malaysia, this was evident in the adjustments Airbus had to make to its training protocols. Initial programs assumed a direct, confrontational style of feedback, where mistakes were identified and corrected on the spot. This clashed with local norms of preserving face and avoiding public embarrassment. As a result, trainers were compelled to adopt a softer approach, embedding critique in indirect forms of guidance and gradually building trust.

Here, the figure of the cultural broker becomes indispensable. These individuals—often bilingual engineers or managers with cross-cultural experience—acted as interpreters not just of language but of meaning. They translated technical instructions into culturally resonant forms, reframing demands in ways that local workers could absorb without feeling humiliated. In doing so, they performed the very essence of anthropotechnology: mediating between the human, the cultural and the technical (Barbaroux & Le Moigne, 2009). The significance of cultural brokers is often underestimated in official reports. Technology transfer is measured in patents acquired, machines installed or personnel trained. Rarely is attention paid to the quiet labor of those who translate across cultural divides. Yet without them, even the most sophisticated systems can falter. They understood the idioms of both worlds, and their presence helped to diffuse tensions that could have escalated into costly delays.

Anthropotechnology pushes us to rethink what counts as "technology" in the first place. If we take seriously the idea that technical systems are inseparable from the social worlds in which they operate, then cultural mediation is not an optional add-on but a core component of technological design. To ignore it is to risk designing artifacts that remain technically elegant but socially unusable (Wisner, 1995). This recognition has profound implications for how we narrate the history of technology transfer in Malaysia. Too often, the story is told as a tale of passive reception, where local firms merely absorb innovations from abroad. Anthropotechnology instead highlights the creative agency of local actors. They do not simply receive; they reshape, reinterpret and repurpose. The Airbus–CTRM collaboration was not a one-way street but a dynamic process of mutual adjustment.

Seen from this angle, anthropotechnology is not a marginal subfield but a necessary corrective to mainstream accounts of globalization. It asks us to see technology not as a universal tool but as a cultural artifact, one that requires ongoing negotiation at the boundaries of difference.

Conclusion

The Airbus–CTRM case ultimately demonstrates that technology transfer is as much about people as it is about machines. The VR training systems, composite materials and manufacturing processes that moved from Europe to Malaysia were not simply technical artifacts. They were carriers of embedded cultural assumptions about time, authority, learning and responsibility. When those assumptions collided with local practices, the results could be either innovation or breakdown.

From the perspective of anthropotechnology, these moments of friction are not failures but opportunities. They reveal the invisible infrastructures of culture that underpin all technical systems. They also highlight the indispensable role of negotiation in any transfer project. Airbus engineers did not merely teach; they had to learn. CTRM workers did not merely adapt; they had to reshape. What emerged was not a pure copy of European aerospace practices but a hybrid system, marked by the imprint of both traditions.

This hybridity is, in fact, the essence of successful technology transfer. Rather than aiming for perfect replication, projects succeed when they foster creative adaptation. Anthropotechnology provides the conceptual tools to see this clearly. By foregrounding the human and cultural dimensions of technology, it shifts our attention from the abstract language of efficiency and productivity to the lived realities of work, trust and meaning.

The broader implication is that Malaysia's aerospace ambitions cannot be understood solely in terms of technical capacity. They hinge equally on the country's ability to cultivate cultural brokers, to recognize the legitimacy of local practices and to design systems that are socially as well as technically robust. This is not unique to Malaysia. All nations engaged in technology transfer face the same challenge: how to balance fidelity to imported systems with sensitivity to local contexts.

Anthropotechnology offers a methodological reminder: to study technology transfer is to study not just engineering diagrams and production schedules but also gestures, silences and relationships. It is to see the factory floor as a cultural landscape, where identities are negotiated and hierarchies reconfigured. The Airbus–CTRM collaboration illustrates this vividly: a project that began as a technical partnership became, in practice, a cultural encounter. And it is in that encounter, messy, contested and productive, that the true meaning of technology transfer resides.

Methodological Note

This study adopts a mixed-methods design that integrates qualitative and quantitative approaches in a sequential and triangulated manner. The overall logic of the research is depicted in Fig. 3.1, which situates data collection within the broader conceptual architecture. Building on this foundation, Fig. 3.2 presents the conceptual model of technology transfer that guided the inquiry.

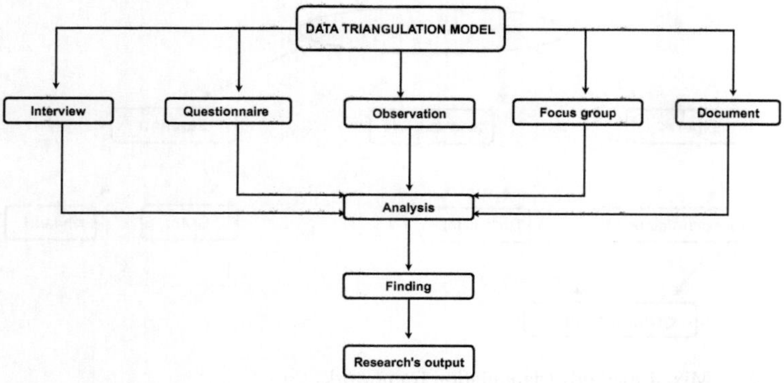

Fig. 3.1 Overall research design of the study

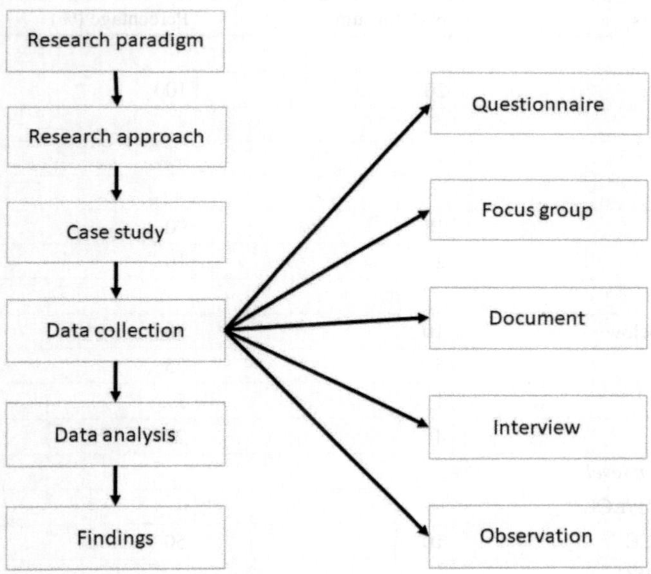

Fig. 3.2 Conceptual model of technology transfer

To ensure robustness, triangulation was employed across sources and methods. Fig. 3.3 illustrates how qualitative and quantitative streams were cross-validated. The empirical material is grounded in a purposive sample of respondents drawn from multiple organizational contexts. Their profile is summarized across Table 3.1, which outline respondent categories and provides a concise overview of the human landscape underpinning the study.

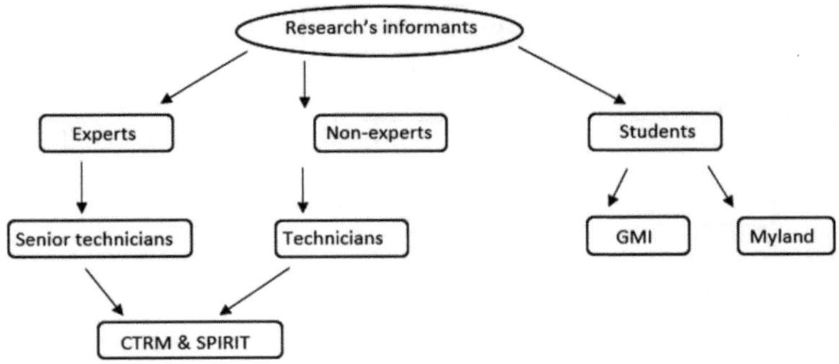

Fig. 3.3 Mixed-methods triangulation framework

Table 3.1 Respondent categories

Categories	N (total sum)	Percentage (%)
Ethnic		
Malay	20	100
Chinese	–	–
Indian	–	–
Gender		
Male	18	90
Female	2	10
Age		
25 and below	10	50
26–30	5	25
31–35	1	5
36–40	4	20
Education level		
PMR/SRP/LCE	–	–
SPM/MCE	10	50
STPM/HSC	–	–
Diploma	10	50
Others	–	–

References

Akrich, M. (1992). The de-scription of technical objects. In W. E. Bijker & J. Law (Eds.), *Shaping technology/building society: Studies in sociotechnical change* (pp. 205–224). MIT Press.

Callon, M. (1986). Some elements of a sociology of translation: Domestication of the scallops and the fishermen of St Brieuc Bay. In J. Law (Ed.), *Power, action and belief: A new sociology of knowledge?* (pp. 196–233). Routledge.

Clegg, C. W., & Corbett, J. M. (1987). Sociotechnical systems. In G. Salvendy (Ed.), *Handbook of human factors* (pp. 833–853). Wiley.

Geertz, C. (1973). *The interpretation of cultures*. Basic Books.

Hutchins, E. (1995). *Cognition in the wild*. MIT Press.

Kapp, S. (2011). Human factors and ergonomics in industrial settings. *Applied Ergonomics, 42*(4), 581–589.

Latour, B. (1996). *Aramis, or the love of technology* (C. Porter, Trans.). Harvard University Press.

Latour, B. (2005). *Reassembling the social: An introduction to actor-network-theory*. Oxford University Press.

Said, N. A. (2024). *A cross cultural technology transfer in Malaysia aerospace industry: An anthropotechnological approach* [Unpublished doctoral dissertation]. Universiti Kebangsaan Malaysia.

Shamsul, A. B. (1996). Nations-of-intent in Malaysia. In S. Tonnesson & H. Antlöv (Eds.), *Asian forms of the nation* (pp. 323–347). Routledge.

Wisner, A. (1985). *Anthropotechnologie: Vers un monde industriel polycentrique*. Octarès Éditions.

Wisner, A. (1997). *Understanding problem building: Ergonomics work analysis*. Taylor & Francis.

Zawawi, I. (2014). Technology, culture and modernity: The Malaysian experience. *Kajian Malaysia, 32*(1), 23–41.

Chapter 4
The VR Tool That Needed Translation

Abstract This chapter examines the cultural dynamics of technology transfer through the case of a VR training tool introduced from Toulouse to CTRM. While the tool was technically sound, its reception highlighted a misalignment between European pedagogical assumptions and Malaysian workplace practices. Ethnographic observation revealed that hesitation, silence and peer-to-peer learning were not failures but culturally meaningful behaviors, reflecting norms of humility, apprenticeship and collective engagement. The chapter explores how adaptation through iterative translation of language, pedagogy and interface design enabled the VR system to resonate with local rhythms, preserving both technical fidelity and worker dignity. By framing breakdowns as diagnostic opportunities, this study underscores anthropotechnology as an ethic of design across cultural difference, where technologies are not neutral artifacts but participants in social negotiation.

Keywords Anthropotechnology · Technology transfer · Cultural translation · Embodied knowledge · Virtual reality training

Introduction

The methodological architecture outlined above provides the foundation for the empirical chapters that follow. We now turn to the detailed case of the VR training tool at CTRM. This chapter examines not only the technical infrastructure of the system but also the cultural encounters that shaped its reception, adaptation and everyday use on the shop floor.

Technology is often imagined as universal. Build it once, test it under rigorous conditions and it should work anywhere. The VR system carried within it a set of assumptions about how learning should occur (assumptions rooted in European pedagogical traditions) that did not align with the rhythms, expectations and sensibilities of the Malaysian shop floor. This chapter examines that misfit. It argues that anthropotechnology is not only a method but an ethic: an approach that recognizes cultural differences as forms of dignity, not deficiencies. Where conventional design

N. Ashikin Said, *Cultural Translation in Technology Design*,
SpringerBriefs in Anthropology, https://doi.org/10.1007/978-3-032-13751-7_4

often treats mismatches as resistance, an anthropotechnological approach reframes them as opportunities for dialogue and redesign.

What happened at CTRM illustrates a larger truth: that technologies do not travel as neutral artifacts. They carry embedded worldviews, pedagogical logics and cultural rhythms. When these encounter new contexts, friction arises. But friction is not failure. It is diagnostic that shows us where translation is needed and how universals can be reworked into particulars. Ethnographic fieldwork on the factory floor revealed that what looked like "failures" were not technical errors but interpretive breakdowns. The VR system treated mistakes as neutral prompts for improvement. Yet for many trainees, error carried social consequences. Across training sessions, younger recruits often responded competitively, turning the VR exercise into a playful contest, comparing scores and teasing each other. Senior workers, however, frowned at this behavior.

One technician remarked: *"Budak-budak ni suka main. Tapi kerja sebenar bukan main-main. Saya rasa macam tak sesuai." ("These youngsters enjoy playing. But real work is not play. I don't think this is appropriate.")* The very features that motivated one group undermined the seriousness of craft for another. The same gamified element carried divergent meanings across generations. Another pattern emerged in peer-to-peer learning. Several trainees confessed they preferred to copy a colleague standing nearby rather than follow the disembodied instructions of the machine: *"Kawan tunjuk lebih senang faham daripada suara mesin." ("It's easier to understand when a friend shows me than when the machine speaks.")*

The VR tool assumed that learning was an individual cognitive process mediated by interface prompts. But the shop floor was organized around apprenticeship, where skills are transmitted socially, by watching, mimicking and practicing together. Taken together, these examples show that breakdowns were not accidents. They were signs. They revealed the cultural logics embedded in the tool and exposed how they clashed with the social fabric of the workplace. Following Lucy Suchman's (1987) insight, plans never fully determine action; meaning emerges in situated practice. The VR tool's plan was precise, but on the factory floor, other rhythms of dignity, humility and apprenticeship asserted themselves.

Embodied Knowledge vs. Simulation

Another tension surfaced between embodied knowledge and simulated performance. The VR system encoded skill as a series of discrete steps: pick up, align, fasten and release. Mastery was measured in terms of accuracy and speed. But for workers, expertise was not just about executing correct sequences. It was about tacit bodily attunement—sensing, adjusting and feeling. One senior technician explained:

"Kalau tangan rasa terlalu licin, itu tandanya resin dah terlebih. Kita kena tahu dengan rasa, bukan dengan mata saja." ("If the hand feels too slippery, that means the resin is too much. You have to know it by feel, not just by sight."). Such sensory attunement could not be captured in the VR simulation. Another trainer observed

a paradox: a trainee who achieved the highest VR score struggled on the actual composite lay-up line.

"Dia pandai ikut komputer. Tapi bila pegang material sebenar, tangan dia tak hidup." ("He's good at following the computer. But when he handles the real material, his hands are not alive.")

The phrase "hands not alive" captured the gap between mechanical replication and embodied mastery. In the world of the factory, skill resides as much in muscle memory and bodily rhythm as in conscious knowledge. Michael Polanyi's (1966) classic distinction between *knowing-that* and *knowing-how* was starkly evident here. The VR tool privileged explicit procedural correctness. Workers relied on tacit, embodied feel. These epistemological divides echo wider critiques of simulation. As Mauss (1934) argued in his essay on "techniques of the body," every culture carries distinctive repertoires of movement, rhythm and posture.

To simulate skill without embedding those bodily techniques is to flatten the richness of embodied practice. Interestingly, small adjustments to the VR training began to address this. Trainers proposed aligning voice prompts with breathing cycles, softening the machine's tone and widening the time allowed for each step. Pauses, once penalized, were revalued as part of careful craftsmanship. Workers began to feel less pressured and performance improved. In these shifts, the interface was reshaped not to erase embodiment but to accommodate it.

Translation in Practice

If the VR tool initially failed to align, the process of adaptation became a form of translation—not only linguistic but structural and ethical. One crucial modification was to reintroduce collective learning. Paired training sessions allowed peers to scaffold each other. Observing and quietly assisting a partner mirrored apprenticeship patterns. Trainers noted that anxiety dropped when workers could learn together. Even the language of the system was reworked. Instead of bureaucratic Malay translations lifted from manuals, prompts were rewritten in colloquial forms, reflecting everyday shop-floor speech. Workers said the new phrasing felt "familiar" and "easier to follow."

Humor, once directed at stiff or awkward translations, shifted toward playful engagement with the system itself—a sign of domestication. These adaptations did not alter the hardware. They reframed the pedagogy. Translation here meant more than swapping words; it meant redesigning interaction to align with cultural logics. It was an iterative, dialogical process, echoing Donald Schön's (1983) idea of "reflection-in-action." Workers were not passive recipients but co-designers, shaping how the tool could live in their environment.

Ethics of Design Across Difference

What does this case teach us about design across cultural boundaries? First, it dismantles the myth of universality. The VR system was not neutral; it carried embedded pedagogies. Second, it demonstrates that misalignments are not simply user deficiencies but structural mismatches. To dismiss them as "language barriers" or "passivity" is to miss the point. Workers were not resisting; they were interpreting through their own logics.

This reframing points toward an ethic of design. Anthropotechnology insists that culture be treated as a resource, not a barrier (Rabardel, 1995; Wisner, 1979). Translation is not a one-off gesture but an ongoing negotiation. The goal is not assimilation as to force users to adapt wholesale, nor it is an abandonment to declare the tool unworkable. It is mediation. At CTRM, this ethic emerged in practice: in softer feedback, in paired learning, in colloquial phrasing. Each adjustment preserved dignity, respected rhythms and honored embodied learning. Design, in this sense, became a moral practice. As Latour (2005) reminds us, technologies are not mere tools; they are actors in networks. How they are shaped and reshaped has consequences for human flourishing.

From Breakdown to Islet

After months of iteration, the VR tool at CTRM no longer resembled the universal system shipped from Europe. Its hardware remained the same, but its meaning had shifted. Silence was reinterpreted as respect, hesitation as precision and mimicry as pedagogy. The system had been translated, localized and domesticated. It was no longer a European export but a Malaysian system.

This process highlights the limits of plug-and-play. Technologies do not cross borders as blank slates. They carry embedded assumptions that must be confronted. But when breakdowns are approached anthropotechnologically, they become breakthroughs. They create opportunities to listen, redesign and co-create. What emerged at CTRM was more than a functioning training tool. It was, in embryonic form, what I later came to call an anthropotechnological islet: a cultural docking point where foreign technologies pause, adapt and become livable. The islet is not grand or universal. It is small, situated and provisional. But without it, transfer falters. With it, translation becomes possible.

Conclusion

The story of the VR tool at CTRM reminds us that innovation is not just about engineering excellence. It is about ethical design across difference. To treat breakdowns as signs of user weakness is to miss their diagnostic power. To ignore embodied knowledge is to flatten skill into empty gestures. To insist on universality is to risk irrelevance. Anthropotechnology offers another way: humility, empathy and co-creation. When technologies meet cultures, they must learn to bend, to listen and to translate. Only then can they take root.

In Malaysia's aerospace sector, the VR system became more than a technical device. It became a site of negotiation, where European logics met Malaysian rhythms and where dignity was preserved alongside efficiency. That negotiation is not a detour. It is the heart of technology transfer. The next chapter turns more explicitly to the voices of workers themselves—their stories, frustrations and insights. If Chapter 4 has traced the frictions of translation, Chap. 5 listens directly to those who lived it, asking: what did it feel like to use a tool that was at once alien and familiar, frustrating and fascinating?

References

Latour, B. (2005). *Reassembling the social: An introduction to actor-network-theory*. Oxford University Press.

Mauss, M. (1934). Techniques of the body. *Economy and Society, 2*(1), 70–88.

Polanyi, M. (1966). *The tacit dimension*. Doubleday.

Rabardel, P. (1995). *Les hommes et les technologies: Approche cognitive des instruments contemporains*. Armand Colin.

Said, N. A. (2024). *A cross cultural technology transfer in Malaysia aerospace industry: An anthropotechnological approach* [Unpublished doctoral dissertation]. Universiti Kebangsaan Malaysia.

Schön, D. (1983). *The reflective practitioner: How professionals think in action*. Basic Books.

Suchman, L. (1987). *Plans and situated actions: The problem of human-machine communication*. Cambridge University Press.

Wisner, A. (1979). *Travail et connaissance*. Editions Sociales.

Chapter 5
Life on the Shop Floor

Abstract This chapter repositions workers' voices as central to the practice and ethics of technology transfer. Too often, global innovation discourses foreground budgets, hardware and managerial strategy while relegating workers' perspectives to footnotes or performance metrics. Yet in CTRM's adoption of Airbus-linked VR training modules, it was precisely the voices of technicians, operators and supervisors that revealed the hidden logics of success and failure. Through ethnographic attention to sound, silence, gesture and storytelling, the chapter shows how Malaysian shop-floor culture reshaped the very meaning of training. Silence functioned as deference rather than confusion; tactile apprenticeship outperformed purely visual sequencing; and informal peer discussions sustained learning beyond formal sessions. These insights challenge universalist design assumptions, demonstrating that transfer is never technical alone but always cultural. By listening, VR was reframed from a rigid imported system into a negotiable, adaptive tool. Anchored in Wisner's anthropotechnology, the chapter argues that listening is not a methodological add-on but an infrastructural act: a means of translating global systems into locally meaningful practice. In this reframing, worker voices become not data points but design inputs—transforming technology transfer into technology transformation.

Keywords Technology transfer · Worker voices · Anthropotechnology · Ethnography · Cultural adaptation

Introduction

When discussions of technology transfer appear in policy documents or corporate strategies, they tend to foreground machines, budgets and expertise. Rarely do they begin with voices. Workers—the very people tasked with making a foreign system operational—often appear only at the margins, their perspectives folded into productivity metrics or training attendance rates. Yet in the case of CTRM's attempt to adopt VR training modules, it was the voices of technicians, operators and supervisors that revealed the true contours of success and failure.

N. Ashikin Said, *Cultural Translation in Technology Design*,
SpringerBriefs in Anthropology, https://doi.org/10.1007/978-3-032-13751-7_5

This chapter argues that listening to workers is not a peripheral activity but a methodological and ethical cornerstone of anthropotechnology. By centering their voices, we see how global technologies are translated, resisted and re-imagined in local shop-floor cultures. The chapter unfolds through thick ethnographic description of the CTRM factory environment; the smell of resin, the rhythm of sanding wings and the subtle gestures exchanged among workers. It is in these seemingly mundane details that broader truths emerge: about hierarchy, about knowledge transmission and about how technology embeds itself into everyday practice. Following Wisner's (1985) principle that technology must adapt to humans and not the reverse, the chapter treats listening as an act of design. It is a practice of making visible the invisible labor, tacit skills and cultural codes that shape whether VR training becomes a bridge or a barrier.

Listening as Method

To listen in a factory is to do more than record. It requires slowing down in a space designed for speed, where parts move along timelines and workers' bodies move in repetitive synchrony. As Clifford Geertz (1973) reminded us, ethnography is about "thick description." The researcher does not simply hear words but interprets pauses, laughs and silences.

In the CTRM shop floor, listening meant leaning into conversations during tea breaks, noting the cadences of Bahasa Malaysia of different dialects, interspersed with technical jargon borrowed from English. It meant following the worker who gestured rather than spoke, tracing how his hands enacted knowledge that no manual contained. This method foregrounds silence as much as speech. As Ong (1987) has shown in her study of Malaysian electronics factories, silence in the workplace is not emptiness but a sign of discipline, deference or sometimes quiet resistance.

At CTRM, the ethnographer's challenge was to distinguish between these registers—to discern when silence meant comprehension, when it meant confusion and when it meant disagreement too risky to voice aloud. Listening, then, was a form of translation. It bridged the highly formalized discourses of Airbus engineers and the grounded, embodied practices of CTRM's technicians. In that translation, ethnography became not only a mode of data collection but a form of collaboration.

Ethnography on the Shop Floor

Stepping into the composite wing assembly hall at CTRM, one was greeted by an atmosphere that was neither entirely industrial nor entirely artisanal. The space smelled of epoxy resin—sharp, chemical and slightly sweet. Large fans hummed in the corners, circulating humid air that clung to the skin. The floor was marked with yellow safety lines, but workers often stepped across them in small improvisations dictated by the rhythm of production.

At one station, two men bent over a mold, their gloved hands smoothing carbon-fiber sheets in layers. The process looked mechanical, but their movements revealed deep tacit knowledge: a flick of the wrist here, a pause there, a subtle nod to indicate readiness for the next step. They rarely spoke. Instead, their communication unfolded through rhythm and gesture. In this environment, the arrival of VR headsets seemed almost alien. Sleek, black devices designed in European labs promised to replicate tasks virtually. But the sensory environment of the shop floor—the smell of resin, the feel of vibration, the background chatter—could not be compressed into pixels and code.

Workers voiced this gap not always directly but through subtle commentary: *"In VR, the part looks clean. But here, look at my gloves. You can't feel the stickiness." "The headset shows one way. But sometimes we adjust — because the real material doesn't behave the same."* Such remarks revealed more than dissatisfaction. They pointed to a disjuncture between embodied, situated knowledge and the abstracted sequences presented in VR.

Workers' Stories

The voices of workers gave the project its most important insights. Rahman, a technician with ten years of experience, described how he first learned to drill composite panels: *"My supervisor didn't just explain with words. He stood beside me, placed his hand on mine, and guided the drill into the panel. I felt the angle, the pressure, the vibration of the tool. Only then did I try it on my own. That is how we learn. VR can guide my hand with some force, but it cannot give me the real vibration."*

A younger recruit, still in his second year, added: *"In VR I can see the hole and even feel some resistance, but I cannot tell if the drill is straight from touch alone. In the workshop, the sound and the vibration of the panel tell me if I'm wrong. The headset is silent. It doesn't warn me in the same way."* Another worker put it more bluntly: *"Drilling is about feel. The machine shows pictures and gives some pushback, but it doesn't teach the hands and ears to work together. That only comes from the real drill."*

Their story emphasized apprenticeship as tactile transmission. VR modules, by contrast, assumed cognition as primarily visual and sequential. Aisyah, one of the few female operators, reflected on hierarchy and silence: *"When the trainer asks questions in the room, many of us keep quiet. It is not that we don't know. We don't want to interrupt, or maybe our English is not smooth. But after the session, we talk among ourselves. Then we understand."*

Her account reframed silence not as ignorance but as cultural etiquette. It also revealed how learning often migrated outside official spaces—into informal peer discussions during breaks. Mani, a line leader, highlighted the mismatch of temporalities: *"In training, they want us to follow each step exactly. But on the line, we sometimes change the order to save time. It is not wrong — it is efficiency. But VR doesn't allow that. It forces us back into their sequence."*

Mani's frustration pointed to a clash between rigid design logic and flexible shop-floor improvisation. Together, these stories illustrate what Burawoy (1979) called the "manufacturing consent" of industrial workplaces: workers comply with systems, but not passively. They reinterpret, adjust and sometimes resist. Their voices are not just commentary but active interventions in the life of technology.

Interpreting Silences

Perhaps the hardest ethnographic task was to interpret moments when workers said nothing. During one VR pilot session, a group of six trainees stood quietly as the instructor demonstrated. Questions were invited; none were asked. An Airbus engineer later whispered: "They don't understand." But when asked afterward in Bahasa Malaysia, the workers explained: *"We understood. But if we ask, maybe the engineer thinks we are stupid. Better to try first."*

This silence was not absence but strategy. It preserved face, maintained hierarchy and protected the collective from embarrassment. In such contexts, silence became a form of communication richer than speech. The implication for training design was profound. Systems that assume constant verbal interaction may misread silence as disengagement. In fact, silence may indicate deep concentration, cultural deference or collective problem-solving that occurs outside the formal setting. As anthropologist James Scott (1990) observed in his analysis of "hidden transcripts," subaltern groups often communicate resistance and negotiation in ways that escape the notice of those in power. At CTRM, silence itself was a hidden transcript—one that only careful listening could decode.

Rethinking Training Through Voices

By centering worker voices, we began to rethink what training should look like. Instead of treating VR as a replacement for embodied apprenticeship, it became clearer that VR could serve as a complement. For instance, VR modules could introduce the general sequence of tasks, while shop-floor mentors filled in the tactile, sensory and tacit dimensions. This insight aligns with Suchman's (1987) argument that plans are resources for situated action, not blueprints to be followed blindly. VR sequences should not dictate reality but provide scaffolds for improvisation. Workers' narratives demonstrated how training succeeds when it acknowledges the legitimacy of local practice.

Pragmatically, this meant redesigning the VR interface to allow "pause and consult" functions, where trainees could check procedures against team discussion. It also meant incorporating narrative prompts, where tasks were framed as part of a collective workflow rather than isolated individual steps. Listening, therefore, was not passive. It actively reshaped the tool. Worker voices became inputs into design—transforming VR from a foreign imposition into a collaborative artifact.

Conclusion

This chapter has argued that technology transfer cannot be reduced to the movement of code, hardware or manuals. At its core, it is about the circulation of meaning. And meaning resides in voices. Listening to workers at CTRM revealed not just gaps in VR design but deeper truths about how knowledge circulates in industrial settings. Workers learn through touch, through rhythm and through silence. They innovate not by rejecting systems but by subtly bending them to fit local realities. To honor these voices is both ethically necessary and practically effective. Without them, VR training would have remained a foreign curiosity. With them, it began to evolve into something usable, adaptable and meaningful. In the broader landscape of global technology transfer, this lesson is urgent.

If we want technologies to land, we must build harbors not just of infrastructure but of empathy. We must listen because in the cadence of worker voices lies the blueprint for a more human-centered future of innovation. The VR tool was not rejected because it was foreign. It was reshaped because workers insisted on making it meaningful. Their improvisations allowed the system to dock in Malaysian soil. This chapter, then, is not just about a training tool. It is about recognizing workers as epistemic agents, as bearers of local wisdom and as co-designers in the global circulation of technology. It is about listening; not as an ethical add-on, but as a practical necessity. Without that listening, tools falter. With it, transfer becomes transformation.

References

Burawoy, M. (1979). *Manufacturing consent: Changes in the labor process under monopoly capitalism*. University of Chicago Press.

Geertz, C. (1973). *The interpretation of cultures*. Basic Books.

Ong, A. (1987). *Spirits of resistance and capitalist discipline: Factory women in Malaysia*. SUNY Press.

Scott, J. C. (1990). *Domination and the arts of resistance: Hidden transcripts*. Yale University Press.

Suchman, L. (1987). *Plans and situated actions: The problem of human-machine communication*. Cambridge University Press.

Wisner, A. (1985). *Ergonomie et anthropotechnologie*. CNRS Éditions.

Chapter 6
Perspective Taking and Local Wisdom

Abstract This chapter explores how cognitive ability and cultural practice intersect in the transfer of aerospace training technologies. At the heart of the study is the Perspective Taking Ability (PTA) test, a computerized measure of spatial cognition, used alongside ethnographic observation of technicians learning to drill composite panels through VR simulations. The findings reveal that PTA scores reflecting egocentric spatial skills, strongly predict performance outcomes such as drilling accuracy and control, particularly for novices. In contrast, allocentric tasks such as Array Rotation (AR) proved less relevant to the shop-floor realities of training. Yet numbers alone could not explain the story. Workers described learning not as solitary mental rotation, but as embodied apprenticeship: feeling pressure on the drill, synchronizing movements with colleagues, narrating tasks through stories and pausing in ways that signaled precision and respect. What appeared to outsiders as inefficiency was, in practice, a rhythm of trust and collective learning. The VR system, designed in a European context, initially struggled to account for these logics. Over time, however, the tool was re-coded. Its prompts softened, its interfaces redesigned and its sequences adapted to resonate with local learning cultures. The lesson is that technology transfer is never a one-way movement of machines or code. It is a process of translation—cognitive, cultural and relational. By combining experimental tests with ethnographic listening, this chapter shows how anthropotechnology helps reveal the hidden intelligence of work, turning apparent failures into insights for design.

Keywords Anthropotechnology · Spatial cognition · VR training · Technology transfer · Apprenticeship

Introduction

Why do some training tools, engineered with precision, falter when introduced into new contexts? The Airbus–CTRM collaboration brought with it not only advanced machinery and software but also a sophisticated VR training system. From a technical standpoint, the system was flawless: error-detection algorithms were precise,

N. Ashikin Said, *Cultural Translation in Technology Design*,
SpringerBriefs in Anthropology, https://doi.org/10.1007/978-3-032-13751-7_6

sequences ran smoothly and simulations replicated physical operations with impressive fidelity. Yet, as earlier chapters have shown, technicians often hesitated, misinterpreted prompts, or struggled to transfer skills from virtual training to actual drilling on composite panels.

This paradox raised a deeper question. Was the challenge rooted in the technology itself or in the way knowledge was structured, embodied and enacted in the Malaysian context? To probe this issue, the research adopted an unusual methodological combination: computerized cognitive testing and paired with ethnographic observation. Specifically, the study employed a PTA task developed by Hegarty and Waller (2004) and programmed by Prof. Maria Kochevnikov for medical training contexts, which was here adapted to the industrial setting of aerospace technicians.

The rationale was straightforward. Drilling composite panels requires highly specialized forms of spatial reasoning. Technicians must align drill bits at precise angles, regulate depth and force and adjust perspective continuously as they move across the work surface. Errors are not merely costly in material terms but can compromise structural integrity and safety. The hypothesis was that differences in egocentric spatial ability (anchored in one's own embodied perspective) versus allocentric spatial ability (imagining from an external viewpoint) might explain variations in drilling performance.

By correlating PTA test outcomes with drilling accuracy and observing training practices ethnographically, the chapter aims to uncover the cognitive and cultural dimensions of skill acquisition. What emerged was a richer picture than numbers alone could capture. Cognition, it turned out, was inseparable from culture, practice and social meaning. What at first appeared to be "failure" of ability was instead evidence of a different orientation to space and learning—one rooted in collective rhythm, narrative memory and embodied apprenticeship.

The PTA Test in Context

The PTA task used in this study was a computerized spatial cognition assessment based on the Hegarty and Waller (2004) paradigm and implemented in software by Prof. Maria Kochevnikov in her earlier research on spatial ability in medical training. The test requires participants to interpret spatial arrangements from imagined viewpoints, rotating objects mentally and responding from alternative perspectives. In this project, we applied the same computerized version directly to aerospace technicians in Malaysia, without modification, to examine how their spatial cognition related to drilling performance and VR training outcomes.

Unlike traditional pen-and-paper versions, this digital test presented technicians with schematic layouts on a computer screen. Participants were asked to imagine themselves standing at a given point, facing a particular direction and then to indicate the location of an object from that perspective.

Each trial generated three measurable outputs:

1. PTA Score: the number of correct responses, reflecting efficiency.
2. Angular Deviation: the difference (in degrees) between the correct orientation and the chosen orientation, reflecting accuracy.
3. Response Time: the time taken (in seconds) to answer, reflecting processing speed.

To complement PTA, we also employed an Array Rotation (AR) test, designed to measure allocentric spatial ability; the capacity to mentally rotate an array of objects from a detached viewpoint. Together, these tasks differentiated between egocentric (self-anchored) and allocentric (world-anchored) cognitive strategies. Why was this relevant to drilling? On the shop floor, technicians constantly shift between perspectives. A driller may lean over a panel, sighting the angle of the bit with one eye, then adjust posture to check perpendicularity. Successful drilling demands fine egocentric control: aligning one's body with the drill, feeling vibration, gauging force. Allocentric reasoning plays a role too.

For instance, when visualizing how holes must line up across a sequence of panels, but the decisive movements are performed from the driller's own embodied perspective. The computerized PTA test thus offered more than abstract psychometrics. It provided a quantifiable proxy for how technicians perceive, orient and act in space. By correlating test scores with drilling outcomes - overshoot (D), time to complete a hole (t), chamfering speed (S) and lateral displacement (L), we could trace how cognitive orientation translated into material performance.

What the Numbers Say

The dataset covered 26 technicians: 10 novices and 16 experts. Each completed both PTA and AR tests, and their results were compared with drilling performance.

Key findings include:

1. PTA predicts performance: PTA scores correlated significantly with drilling outcomes, especially overshoot (D) and displacement (L).
2. AR does not predict performance: No significant correlations were observed between AR (allocentric ability) and drilling performance.
3. Novices vs. experts: For novices, PTA strongly predicted performance (negative correlation with overshoot and displacement), while for experts correlations were weaker or absent.
4. Training effects: Novices improved speed but often sacrificed accuracy, suggesting overconfidence where they thought they were faster but actually produced more errors. Experts, by contrast, balanced efficiency with accuracy, showing stable egocentric control.

Summary of findings are compressed in Table 6.1.

The PTA score is an efficiency measure, capturing both speed and correctness. Angular deviation reflects accuracy, i.e., the average difference between correct and chosen orientations. The pattern was clear:

Table 6.1 Correlations between cognitive tasks and drilling performance

Group	Significant PTA correlations	AR correlations	Interpretation
Novices	PTA score negatively correlated with overshoot ($r = -0.79, p < 0.01$) and displacement ($r = -0.73, p < 0.05$)	None	Egocentric ability is crucial for learning precision; allocentric irrelevant
Experts	Weak or no significant correlations	None	Expertise stabilizes performance regardless of test scores
Overall	PTA correlates with drilling outcomes (overshoot, displacement)	None	Confirms predictive value of egocentric perspective

1. Novices became faster but less accurate. They often overshot or misplaced the drill tip, leading to rework. This reflected overconfidence.
2. Experts drilled at consistent speeds but with greater precision. Overshoot was rare and displacement minimized.

Thus, PTA predicted not only immediate outcomes but also the trajectory of skill acquisition: novices who scored higher on PTA were more likely to improve drilling performance, while those with lower scores plateaued.

Beyond the Numbers: Local Wisdom

The test data were striking, but they made full sense only when paired with ethnographic observation:

1. Narrative memory: One technician narrated tasks as stories rather than following sequential steps. Trainers initially dismissed this as digression, but in practice, it encoded memory more effectively than rote listing.
2. Collective rhythm: Another worker paused three times before drilling. The VR flagged inefficiency, but colleagues recognized it as rhythm—syncing with team flow.
3. Embodied trust: Silence during training, often misread as hesitation, actually signaled attentiveness and respect.

These practices revealed intelligence expressed in registers invisible to the VR system. As Masuda and Nisbett (2001) argue, East Asian cognition often attends more to relational and contextual cues than to focal objects. The Malaysian technicians' orientation was consistent with this broader cultural pattern. Here, anthropotechnology played a critical role: reframing what seemed like inefficiency as alternative modes of knowing. Instead of asking why workers "failed" to adapt to VR, we asked what the VR failed to see in them (Simondon, 2017). Armed with these insights, our design approach shifted. We redesigned VR modules to ground tasks in familiar contexts:

1. Realistic layouts: Virtual environments mirrored actual workshop settings.
2. Story-driven prompts: Instructions became narrative-based, encouraging technicians to "tell the drill" as they worked.
3. Collaborative checkpoints: Pause-and-ask features acknowledged that learning often unfolded collectively.

The result was palpable. Performance improved, missteps decreased and engagement rose. Workers began to see the VR system not as foreign imposition but as an extension of their craft. This was more than localization as surface decoration; it was cognitive redesign. It echoed STS arguments that technologies are always translated as they move across contexts (Akrich, 1992; Latour, 1987). The PTA test revealed something fundamental: perspective-taking is not universal but situated. Designing effective training required more than identifying deficits; it required seeing with workers rather than at them.

This resonates with Lave and Wenger's (1991) notion of situated learning—knowledge embedded in practice rather than abstracted as universal form. It also echoes Geertz's (1973) warning against assuming universality where difference prevails. In practical terms, this meant valuing narrative, rhythm and silence as legitimate cognitive strategies. In epistemological terms, it meant recognizing that cognition is always cultural, and training must be designed as translation, not imposition.

The juxtaposition of test results and ethnographic insights reframed cognition as situated, relational and cultural.

1. Situated cognition: Skills were enacted in practice, not abstracted from it (Lave & Wenger, 1991).
2. Cultural cognition: As Masuda and Nisbett (2001) argue, East Asian cognition emphasizes relational cues; Malaysian technicians embedded drilling in rhythm and story, not abstract object focus.
3. Design implications: VR modules were redesigned to reflect local logics: realistic layouts, narrative prompts and collaborative checkpoints.

Performance improved measurably, but more importantly, the VR tool was no longer seen as an imposition. It became part of technicians' apprenticeship ecology.

Conclusion

Looking back, what seemed at first like inefficiency was, in fact, intelligence. The pause was precision. The narrative was encoding. The rhythm was trust. The PTA test helped reveal cognitive roots of these practices, but it was daily observation of listening to workers and watching gestures that made them meaningful.

The lesson is clear: in cross-cultural training, tools must do more than transfer skills. They must learn to translate sense-making. Sometimes, the best innovation does not teach something new. It reveals the wisdom already there. For CTRM, the PTA test was more than a cognitive instrument; it was a mirror. It showed that perspective is not only a spatial skill but also a cultural practice. And in that recognition lies the future of anthropotechnology: designing not for universal humans, but for specific worlds.

References

Akrich, M. (1992). The de-scription of technical objects. In W. Bijker & J. Law (Eds.), *Shaping technology/building society* (pp. 205–224). MIT Press.

Geertz, C. (1973). *The interpretation of cultures*. Basic Books.

Hegarty, M., & Waller, D. (2004). A dissociation between mental rotation and perspective-taking spatial abilities. *Intelligence, 32*(2), 175–191.

Lave, J., & Wenger, E. (1991). *Situated learning: Legitimate peripheral participation*. Cambridge University Press.

Latour, B. (1987). *Science in action: How to follow scientists and engineers through society*. Harvard University Press.

Masuda, T., & Nisbett, R. E. (2001). Attending holistically versus analytically: Comparing the context sensitivity of Japanese and Americans. *Journal of Personality and Social Psychology, 81*(5), 922–934.

Simondon, G. (2017). *On the mode of existence of technical objects*. Univocal.

Chapter 7
Collaboration in Translation: Negotiating Technology and Culture

Abstract This chapter explores the dynamics of technology transfer in the aerospace sector through the case of the VIRISTAM project, which involved the adaptation of a VR training tool originally developed by Airbus for application within Malaysia's CTRM. By examining the transfer of advanced simulation technology, the chapter highlights the intersecting roles of cultural adaptation, institutional frameworks and collaborative learning across organizational and national boundaries. Using the Triple Helix model of university–industry–government relations as a conceptual anchor, the study demonstrates how the VIRISTAM initiative not only facilitated technological upgrading but also revealed the sociocultural dimensions that underpin successful knowledge transfer. Particular emphasis is given to the role of cultural brokers—mediators who navigate between differing organizational logics and cultural expectations—to ensure that imported technologies resonate with local practices and values. Through qualitative data and narrative reconstruction, the chapter illustrates how such brokerage enables both the localization of global technologies and the globalization of local expertise. The discussion underscores that technology transfer is never a purely technical process; it is equally shaped by negotiations of identity, institutional trust and cultural meaning. The VIRISTAM case thereby contributes to a broader understanding of how emerging economies can strategically position themselves in global value chains by not only acquiring technological tools but also embedding them within their own socio-institutional landscapes.

Keywords Technology transfer · Cultural brokers · Triple Helix · VR training · VIRISTAM

Introduction

Collaboration is often spoken of as a precondition for successful innovation, a resource to be harnessed, or a structure to be formalized through agreements and institutions. In the language of policy documents, collaboration appears as a neat architecture in which universities, industries and governments work together within

N. Ashikin Said, *Cultural Translation in Technology Design*,
SpringerBriefs in Anthropology, https://doi.org/10.1007/978-3-032-13751-7_7

the Triple Helix model (Etzkowitz & Leydesdorff, 2000). Yet, when one moves from the polished abstractions of innovation policy to the lived practice of collaboration, the picture becomes considerably more complex. This chapter examines collaboration not as a background condition but as the central work of technology transfer, drawing from the case of the VIRISTAM project, which sought to adapt a VR training tool developed by Airbus for use by CTRM.

At the surface, the partnership appeared straightforward: Airbus provided the technological expertise; CTRM offered industrial application and local workforce; the AMIC facilitated software, technological needs and coordination; and Malaysian universities, particularly Universiti Kebangsaan Malaysia (UKM) and Universiti Malaya (UM), contributed academic insights. Together, they formed what policymakers heralded as an exemplary Triple Helix partnership, embodying the synergy of state, industry and academia in driving Malaysia's aerospace ambitions. But beneath this narrative of alignment, collaboration unfolded as a fragile, ethically demanding and often unpredictable process.

This chapter argues that in the VIRISTAM case, collaboration was not a structural arrangement but an ongoing ethical labor, involving the translation of technical knowledge across cultural, institutional and disciplinary domains. It was marked by moments of silence, hesitation and tacit negotiation that carried as much weight as formal agreements. By tracing the microdynamics of this collaboration, the chapter reconsiders technology transfer not as a linear movement of tools from one context to another but as a process of collective world-making, in which diverse actors struggled to recognize one another, build trust and align priorities.

Theoretical Framing: From Triple Helix to Collaboration as Ethical Labor

The Triple Helix model has been one of the most influential frameworks in understanding innovation systems. Developed by Etzkowitz and Leydesdorff (2000), it envisions collaboration as the dynamic interaction of universities, industries and governments in fostering knowledge economies. Its strength lies in offering a conceptual shorthand for describing how institutional spheres intersect and generate synergies. In Malaysia, policymakers have frequently invoked the Triple Helix in articulating national innovation strategies, particularly in high-technology sectors such as aerospace (Govindaraju et al., 2013).

The Triple Helix has often been visualized as three interlocking spirals, each representing university, industry and government, moving in dynamic tension rather than in fixed alignment (Etzkowitz & Leydesdorff, 2000). In the context of VIRISTAM, this model provided a useful heuristic to describe how Airbus, CTRM, Malaysian policy actors and local universities were formally positioned in relation to one another. Yet, as the subsequent analysis will demonstrate, these connections were neither seamless nor automatic; they required constant mediation and translation across institutional boundaries.

Figure 7.1 (author's own) situates the VIRISTAM project within this broader Triple Helix framework, illustrating how each actor was drawn into the collaboration and setting the stage for examining the lived negotiations that sustained the project. Yet, as critics have pointed out, the Triple Helix tends to flatten the complexities of collaboration into structural categories. It presumes alignment of goals, overlooks asymmetries of power, and underplays the ethical and affective labor required to make partnerships function (Ranga & Etzkowitz, 2013). In the VIRISTAM project, while the Triple Helix was present as an official discourse, the lived reality of collaboration demanded constant improvisation and translation.

To grasp this reality, it is useful to draw from the sociology of translation (Callon, 1986; Latour, 1987). Translation here refers not only to linguistic exchange but to

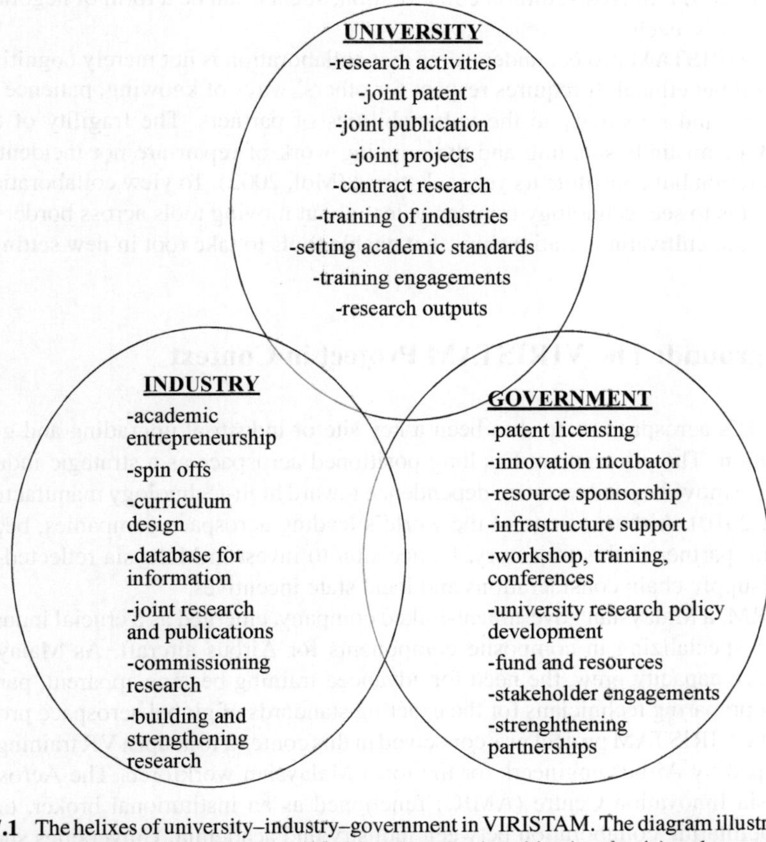

Fig. 7.1 The helixes of university–industry–government in VIRISTAM. The diagram illustrates the interaction between Airbus (industry), CTRM and local universities (academia) and governmental actors in Malaysia's aerospace ecosystem, highlighting the role of cultural brokers in sustaining collaboration. (Author's own 2024)

the active work of rendering ideas, practices and technologies intelligible across different worlds. Collaboration requires more than formal agreements; it involves reconfiguring expectations, negotiating meanings and accommodating forms of tacit knowledge (Polanyi 1966). In this sense, collaboration is not a fixed structure but a process of world-making, in which actors continuously negotiate what counts as relevant knowledge and legitimate practice.

Michael Polanyi's notion of tacit knowledge that "we can know more than we can tell" (1966, p. 4) is particularly salient in the VIRISTAM case. Much of the expertise of Malaysian technicians could not be captured in technical manuals or VR simulations but resided in embodied skills, routines and shared understandings. Recognizing this tacit knowledge was essential for adapting Airbus's VR tool to the Malaysian context. Moreover, silence itself often functioned as a communicative form, signaling hesitation, resistance, or consent. As anthropologists of communication remind us, silence is not the absence of meaning but a mode of meaning-making (Nakane, 2007). In cross-cultural collaboration, silence can be a form of negotiation as potent as speech.

The VIRISTAM project underscores that collaboration is not merely cognitive or structural but ethical. It requires respect for others' ways of knowing, patience with ambiguity and sensitivity to the vulnerabilities of partners. The fragility of trust, the risk of misunderstanding and the ongoing work of repair are not incidental to collaboration but constitute its very substance (Mol, 2002). To view collaboration in this light is to see technology transfer as less about moving tools across borders and more about cultivating relationships that enable tools to take root in new settings.

Background: The VIRISTAM Project in Context

Malaysia's aerospace sector has been a key site of industrial upgrading and global integration. The government has long positioned aerospace as a strategic industry, aiming to move beyond resource dependence toward high-technology manufacturing (MITI, 2010). Airbus, as one of the world's leading aerospace companies, became a central partner in this trajectory. Its decision to invest in Malaysia reflected both global supply chain considerations and local state incentives.

CTRM, a Malaysian government-linked company, emerged as a crucial industrial partner, specializing in composite components for Airbus aircraft. As Malaysia's aerospace capacity grew, the need for advanced training became apparent, particularly in preparing technicians for the exacting standards of global aerospace production. The VIRISTAM project was conceived in this context: to adapt a VR training tool developed by Airbus engineers for the local Malaysian workforce. The Aerospace Malaysia Innovation Centre (AMIC) functioned as an institutional broker, tasked with facilitating collaboration between industry and academia. Universities such as UKM and UM were enlisted to contribute research expertise, including cultural and pedagogical insights into the adaptation of VR training tools. For policymakers, this

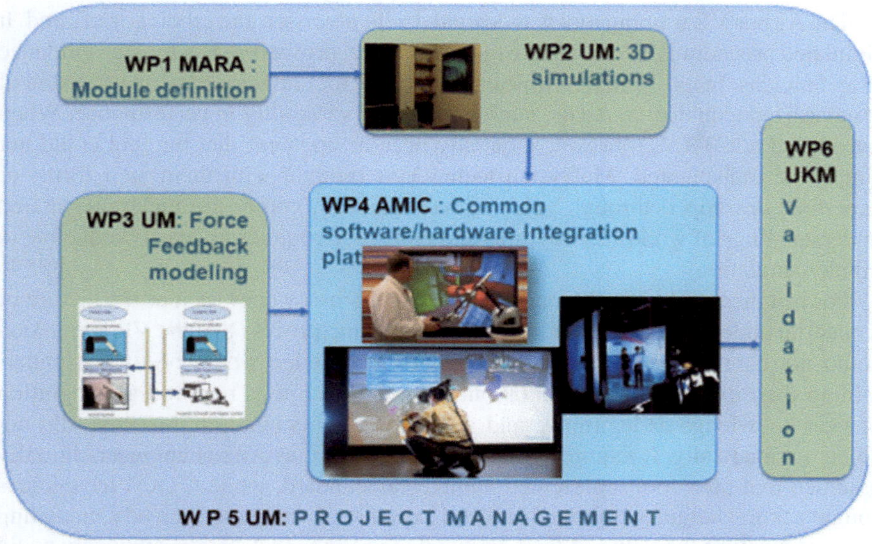

Fig. 7.2 Workflow structure of the VIRISTAM project, showing the division of responsibilities across university partners (UM, UKM, MARA), AMIC, and project management. The diagram highlights the formal role allocation between academic, industrial and coordinating bodies. (Author's own 2024)

was a model Triple Helix project: state agencies provided funding and oversight, industry offered technology and academia contributed knowledge (Fig. 7.2).

In official accounts, the VIRISTAM project was presented as a success story of seamless collaboration. Yet, those involved in the day-to-day work experienced it differently. Misalignments of expectation, differences in institutional cultures and subtle dynamics of recognition complicated the partnership. While Airbus engineers emphasized standardization and efficiency, Malaysian technicians valued contextual appropriateness and tacit practices. Universities often found themselves mediating these divergent priorities, translating not only between languages but between epistemologies.

Collaboration in Practice: Negotiating Knowledge, Silence and Priorities

Collaboration in the VIRISTAM project unfolded in small, often overlooked moments: a pause during a training session, a gesture of hesitation in a meeting or a quiet acknowledgment of local expertise. These moments reveal how the process of technology transfer was sustained less by formal structures than by the everyday labor of negotiation.

The Airbus VR training tool was designed with precision and efficiency in mind. It simulated procedures for composite manufacturing, providing step-by-step guidance to technicians. In its original European context, the tool functioned as an extension of standardized technical manuals, aiming to reduce variability in performance. When introduced at CTRM, however, it quickly became apparent that the tool could not simply be transplanted. Malaysian technicians brought with them tacit forms of expertise; developed through years of working with composite materials, shared understandings of workplace rhythms and embodied routines not easily reducible to digital simulation.

For instance, technicians would often rely on sensory cues such as the texture of a material under their fingers or the sound of a curing process that the VR simulation could not capture. To ignore these forms of knowledge would be to risk undermining their expertise. The adaptation of the VR tool thus required a recognition of tacit knowledge as legitimate and valuable. This recognition, however, did not come automatically. It required persistent negotiation, as Airbus engineers initially assumed that their tool represented a universal standard, while CTRM technicians sought acknowledgment of their localized expertise. Universities played a mediating role, articulating the value of tacit knowledge in terms that Airbus engineers could recognize as scientifically legitimate.

One of the most striking features of collaboration in the VIRISTAM project was the role of silence. In meetings between European engineers and Malaysian technicians, silence often emerged as a form of communication. A technician's quiet pause could indicate disagreement without direct confrontation, reflecting cultural norms of respect and hierarchy (Nakane, 2007). At other times, silence functioned as a strategy of negotiation, allowing space for reflection or signaling the limits of one's willingness to adapt.

Airbus engineers, accustomed to more explicit forms of communication, sometimes interpreted silence as consent or indifference. Misunderstandings followed, requiring careful repair. Over time, however, some engineers came to recognize silence as a meaningful form of participation, one that conveyed caution, resistance, or even tacit critique. Acknowledging silence as communication was itself an ethical gesture, signaling respect for different communicative norms.

Collaboration was further complicated by divergent institutional priorities. For Airbus, the priority was efficiency and standardization; for CTRM, it was contextual appropriateness and the well-being of its workforce; for universities, it was research output and knowledge production; and for the Malaysian state, it was the projection of national capacity in high-technology sectors. These priorities did not always align. At times, they pulled the project in different directions. For example, Airbus pushed for rapid deployment of the VR tool, while CTRM emphasized the need for gradual adaptation to local practices. Universities, meanwhile, sought to document and publish insights from the process, which industry partners sometimes viewed as distractions. The work of collaboration lay in negotiating these tensions without allowing them to fracture the partnership.

Collaboration as Ethical Labor

The VIRISTAM project reveals collaboration as a form of ethical labor. It required ongoing attention to relationships, recognition of differences and repair of misunderstandings. This labor was fragile and often invisible, yet it was what sustained the project.

Trust was not given but built incrementally, through small gestures of recognition. When Airbus engineers acknowledged the tacit expertise of Malaysian technicians, trust deepened. When technicians saw their practices reflected in the adaptation of the VR tool, they felt valued. Conversely, moments of dismissal or misunderstanding could erode trust quickly. Collaboration demanded vigilance in maintaining these fragile bonds. Recognition was central to collaboration. To recognize tacit knowledge, silence, or divergent priorities as legitimate was to affirm the value of partners' contributions. Yet recognition was often partial, uneven, or contested. Misrecognitions occurred, leading to moments of tension. Repair work followed—clarifying misunderstandings, apologizing for dismissals, or adjusting expectations. This repair work, though rarely acknowledged in policy narratives, was essential to keeping collaboration alive.

By foregrounding ethical labor, the VIRISTAM case challenges structural models of collaboration. The Triple Helix provides a useful map, but it cannot capture the lived work of translation, recognition and repair. Collaboration is not a backdrop against which technology transfer unfolds; it is the work itself. It is the fragile fabric that enables technologies to move across borders and take root in new cultural and industrial landscapes.

Conclusion

The VIRISTAM project illustrates the centrality of collaboration in technology transfer. Far from being a simple transfer of tools from one context to another, the process involved the translation of knowledge, the negotiation of silence and the recognition of tacit expertise. Collaboration was not a structural arrangement but an ongoing ethical labor, sustained by fragile trust and continual repair. This case invites a rethinking of how innovation and technology transfer are conceptualized. Rather than viewing collaboration as a stable structure linking universities, industries, and governments, it should be understood as a dynamic and ethically charged practice.

It is in the small, everyday moments of negotiation—pauses, silences, acknowledgments—that the future of technology transfer is shaped. For policymakers, this suggests that fostering innovation requires more than building institutions or funding partnerships. It requires cultivating the conditions for ethical collaboration: spaces where tacit knowledge is valued, silence is respected and divergent priorities can be negotiated. For scholars, the VIRISTAM case underscores the importance of attending to the microdynamics of collaboration, where the real work of technology transfer takes place. In the end, collaboration is not the backdrop to innovation. It is the fragile fabric of innovation itself.

References

Callon, M. (1986). Some elements of a sociology of translation: Domestication of the scallops and the fishermen of St. Brieuc Bay. In J. Law (Ed.), *Power, action and belief: A new sociology of knowledge?* (pp. 196–233). Routledge.

Etzkowitz, H., & Leydesdorff, L. (2000). The dynamics of innovation: From national systems and "Mode 2" to a Triple Helix of university–industry–government relations. *Research Policy, 29*(2), 109–123.

Govindaraju, V. G. R. C., Rasiah, R., & Ramanathan, S. (2013). University–industry collaboration in the Malaysian context. *Science, Technology and Society, 18*(1), 1–25.

Latour, B. (1987). *Science in action: How to follow scientists and engineers through society*. Harvard University Press.

MITI. (2010). *Third Industrial Master Plan (IMP3): 2006–2020*. Ministry of International Trade and Industry.

Mol, A. (2002). *The body multiple: Ontology in medical practice*. Duke University Press.

Nakane, I. (2007). *Silence in intercultural communication: Perceptions and performance*. John Benjamins.

Polanyi, M. (1966). *The tacit dimension*. Routledge.

Ranga, M., & Etzkowitz, H. (2013). Triple Helix systems: An analytical framework for innovation policy and practice in the knowledge society. *Industry and Higher Education, 27*(3), 237–262.

Chapter 8
What Makes Technology Transfer Work (or Fail)

Abstract Technology transfer is often described as a technical and managerial exercise, where innovations designed in one setting are simply deployed elsewhere. Yet my experience with Airbus's VR training tool at CTRM in Malaysia revealed a more complex reality. The system did not fail because of hardware malfunction or software error; it faltered because its design carried cultural assumptions that misaligned with local work practices. Through immersive observation, dialogue and iterative adaptation, I learned that successful transfer requires more than technical robustness—it demands cultural fit, absorptive capacity and trust. This chapter weaves together personal vignettes from my fieldwork moments of hesitation in the VR lab and side conversations with trainers over teh tarik, with theoretical insights from STS, diffusion of innovations and anthropotechnology. The chapter argues that technology does not transfer on its own; people transfer it through processes of translation, compromise and co-creation. By situating CTRM's experience within broader global cases, the chapter identifies five elements crucial for transfer: contextual research, stakeholder involvement, iterative design, trust infrastructure and cognitive fit. Together, they form a framework for understanding why some transfers succeed while others fail. More importantly, they highlight the ethical responsibility of engineers, managers and policymakers to treat technology transfer not as delivery but as dialogue.

Keywords Technology transfer · Anthropotechnology · Cultural adaptation · Absorptive capacity · Trust

Introduction

Technology transfer is one of globalization's most paradoxical processes. On paper, it appears straightforward: a system is designed in one context, rigorously validated and then "transferred" to another site, where it is expected to function identically. Industrial roadmaps and policy documents frame transfer as delivery, as if technologies were sealed containers to be shipped and unboxed intact. Yet our experience at

N. Ashikin Said, *Cultural Translation in Technology Design*,
SpringerBriefs in Anthropology, https://doi.org/10.1007/978-3-032-13751-7_8

CTRM painted a more complicated picture. The success or failure of transfer rarely hinges on technical sophistication alone; it is profoundly shaped by the human, organizational and cultural systems into which technology lands.

I remember the first day we introduced the VR headsets. Workers were curious but cautious. Trainers exchanged knowing glances. Managers hovered, eager to see efficiency gains. On paper, everything seemed aligned: the tool had been tested, certified and praised by Airbus in Europe. But in practice, the VR module initially sat underutilized and adoption was uneven. It became clear that what "success" looked like in Toulouse did not translate directly to Melaka. What then determines whether technology transfer succeeds? Drawing on both our observations at CTRM and insights from global comparative cases, we came to understand that transfer depends less on the inherent qualities of a tool and more on its alignment with the social fabric of its new environment. STS scholars remind us that technologies are not neutral (Akrich, 1992).

They arrive embedded with assumptions, values and "scripts" about their intended users and the contexts in which they operate. When these scripts resonate with local practices, transfer can flourish; when they clash, breakdowns are inevitable. In Malaysia, we discovered early on that cultural misfit was the primary obstacle. Hierarchical expectations shaped how trainers and workers interacted, revealing that even flawless technical design cannot substitute for cultural fit.

The Myth of Plug-And-Play

One of the most persistent myths in technology transfer is that of plug-and-play. In this imaginary, technology behaves like universal software: design once, deploy anywhere. Its appeal is obvious: efficiency, scalability and predictability. A VR module that succeeds in Toulouse should, logically, succeed in Melaka. A gamified training system that motivates European learners should do the same for Malaysian technicians.

Yet our observations at CTRM told a different story. The VR module was technically flawless; the simulations ran smoothly, the headset responded accurately and metrics recorded progress efficiently. However, initial uptake was patchy. Workers reverted to paper-based instructions, trainers quietly reverted to established pedagogical methods and managers who expected to showcase the technology faced questions from senior leadership about underutilization.

The issue was not mechanical failure; it was cultural misfit. The module assumed that trainees would welcome individual feedback, treat errors as learning opportunities and be motivated by gamification. These assumptions reflected European pedagogical traditions emphasizing individualism, direct engagement with mistakes and linear learning progression. In Malaysia, learning was more collective, public failure carried heavier social weight and respect for hierarchy shaped interactions. Feedback intended as neutral was interpreted as shaming; motivation designed for engagement risked trivialization.

This aligns with Rogers' (2003) diffusion of innovations theory: adoption is determined not solely by technical superiority but by compatibility with existing values and practices. Similarly, Akrich (1992) emphasizes that technologies carry scripts about their users, and failure arises when these scripts clash with local realities. The CTRM experience was far from unique. European agricultural machines in Africa often failed not due to poor engineering but because they assumed large, uniform fields and uninterrupted labor rhythms absent in local contexts (Biggs, 1990). Wisner (1995) conceptualized anthropotechnology precisely to address such misalignments: the realization that human contexts are not interchangeable, and that design assumptions must be adapted to local practices. Plug-and-play works in software updates but collapses in complex social systems.

Absorptive Capacity as an Ecology

If plug-and-play is a myth, what determines meaningful uptake? One key concept is absorptive capacity, coined by Cohen and Levinthal (1990), describing an organization's ability to recognize, assimilate and apply new knowledge. Transfer depends not just on the artifact itself but on whether the receiving environment can make sense of it.

At CTRM, absorptive capacity was less about policies or organizational charts and more about an ecology of readiness distributed unevenly across the organization. Workers were curious but cautious; trainers protective of established rhythms; managers sought efficiency but feared delays; policymakers sought measurable impact but struggled to define what VR success looked like. Younger workers, often gamers themselves, engaged eagerly. Older workers, more experienced but less familiar with immersive technology, hesitated. Trainers who felt their tacit knowledge was overlooked withdrew support, signaling to trainees that VR was optional. Managers whose KPIs were not aligned with VR outcomes allowed enthusiasm to wane. Absorptive capacity, therefore, was a lived dynamic, not an abstract organizational trait.

Tacit knowledge was central. Nonaka and Takeuchi (1995) argue that knowledge creation relies on converting tacit into explicit knowledge. VR sequences risked flattening nuanced, embodied skills into rigid step-by-step instructions. Iterative translation between tacit insights and explicit system design became crucial. For instance, we adjusted the VR voice prompts to align with natural breathing rhythms, slowed sequences to match learning tempo and reintroduced collaborative scaffolding for collective training; small changes with outsized impact. Our reflections at CTRM made it clear that absorptive capacity thrives where curiosity, trust and flexibility intersect. A tool's technical sophistication matters little if the human environment is not prepared to interpret, integrate and adapt it.

Absorptive capacity sounded abstract when I first encountered Cohen and Levinthal (1990). But at CTRM, it came alive. Some workers were eager: "*I play games, so this feels natural,*" a younger technician said, adjusting the headset like a pro. Others were cautious: "*I've done this job for years with my hands. Why must I learn through a screen?*" Trainers were split. Some appreciated the standardization VR promised; others feared being sidelined. Managers, too, measured success differently. For one, it was about speed; for another, about compliance with Airbus standards.

Absorptive capacity, I realized, wasn't just an organizational trait, but it was a living ecology. It depended on curiosity, trust and whether tacit knowledge could be honored. Nonaka and Takeuchi's (1995) ideas about tacit versus explicit knowledge echoed here. The VR tool made knowledge explicit, but technicians' embodied expertise risked being flattened. Only when we slowed the sequences, adjusted the voice prompts to match natural breathing rhythms and reintroduced group-based scaffolding did uptake improve. Absorptive capacity was not about the headset. It was about readiness, adaptation and humility.

Trust and Feedback: The Hidden Infrastructure

If absorptive capacity addresses cognition, trust underpins relational dynamics. Without it, workers would not voice doubts, CTRM would not allow external researchers access and government partners would not extend support beyond the pilot phase. Building trust was iterative and relational. We learned that informal moments of coffee chats, shared lunches and casual walkthroughs of the VR lab were as important as formal workshops. Trust did not eliminate conflict; it created a space where conflict could be aired productively. Trust allowed these observations to become design insights. Workers hesitated to wear headsets; trust enabled us to ask why without fear of blame

Feedback itself was reframed as data, not criticism. Every hesitation, workaround and humorous comment revealed misfits between system logic and lived practice. This approach shifted the tone from judgment to collaboration, accelerating iteration and co-ownership. Anthropotechnology underscores that feedback loops are central to design: just as code must be debugged, cultural assumptions must be tested, adjusted and refined.

Toward a Framework for Fit

From our observations, five recurring elements determined whether transfer took root:

1. Contextual Research: Ethnography, observation and cultural listening are prerequisites. Peer-learning, central to Malaysian training practices, informed VR features that improved engagement.

2. Stakeholder Involvement: Early, genuine participation fosters ownership. Trainers' input on pacing and module design made the system realistic and respected.
3. Iterative Design: Transfer is never a one-shot event. Cycles of feedback and adjustment—from voiceover style to group-based sequencing proved essential.
4. Trust Infrastructure: Beyond contracts, relational trust permits honest critique, uncovering misfits early.
5. Cognitive Fit: Tools must align with local ways of learning, perceiving and acting. Embedding perspective-taking and collective scaffolding enhanced adoption.

Missing any of these elements risks failure; not due to technical flaws, but because the technology remains culturally foreign. Embraer's knowledge transfer in Brazil succeeded because of trust and investment in long-term absorptive capacity (Goldstein, 2002). African e-learning initiatives often faltered because localization was treated as mere translation, not deep cognitive redesign (Unwin, 2005). CTRM's case fits within this global pattern: fit matters more than flash.

Our reflections extended beyond technical and organizational lessons. Technology transfer is an ethical practice. Designers and engineers carry responsibility not only for performance metrics but for human dignity, respect and cultural resonance. At CTRM, ethical practice meant listening attentively, negotiating changes collaboratively and acknowledging the validity of local expertise. From a practical perspective, project teams must cultivate patience, humility and openness. VR modules were more successful not because they were superior, but because we slowed down, observed, and adapted iteratively. Recognizing the limits of universal design fosters solutions that are both technically sound and socially acceptable.

Conclusion

The most profound lesson is epistemological: technology transfer is dialogue, not delivery. Delivery says, "Here is the tool; use it." Dialogue says, "Here is a prototype; let's shape it together." Delivery instructs users to adapt; dialogue encourages understanding. Dialogue is slower, messier and less predictable but it works.

Technology does not travel alone; it carries assumptions, scripts and cultural cargo. Successful transfer requires humility, listening and shared ownership. When technology finally speaks the language of its users; linguistically, socially and cognitively, only then it becomes co-created. In Malaysia, the VR headset evolved from a European export to a Malaysian tool through cycles of adaptation, negotiation and trust. The larger lesson is clear: global technology transfer is not merely a technical exercise but an ethical one. Tools must be reshaped by the human contexts they enter. Technology itself does not transfer; people do.

References

Akrich, M. (1992). The description of technical objects. In W. E. Bijker & J. Law (Eds.), *Shaping technology/building society* (pp. 205–224). MIT Press.

Biggs, S. (1990). A multiple source of innovation model of agricultural research and technology promotion. *World Development, 18*(11), 1481–1499.

Cohen, W. M., & Levinthal, D. A. (1990). Absorptive capacity: A new perspective on learning and innovation. *Administrative Science Quarterly, 35*(1), 128–152.

Goldstein, A. (2002). The political economy of high-tech industries in developing countries: Aerospace in Brazil, Indonesia and South Africa. *Cambridge Journal of Economics, 26*(4), 521–538.

Nonaka, I., & Takeuchi, H. (1995). *The knowledge-creating company: How Japanese companies create the dynamics of innovation.* Oxford University Press.

Rogers, E. M. (2003). *Diffusion of innovations* (5th ed.). Free Press.

Unwin, T. (2005). Towards a framework for the use of ICT in teacher training in Africa. *Open Learning: The Journal of Open, Distance and e-Learning, 20*(2), 113–129.

Wisner, A. (1995). Understanding problems of technical change: Anthropotechnology and work culture. Taylor & Francis.

Chapter 9
Inventing the Anthropotechnological Islet

Abstract Technology transfer is often framed as a technical matter: the seamless movement of a finished system from one site to another. Yet the reality of cross-border transfer is rarely so neat. This chapter introduces the concept of the *Anthropotechnological Islet*: a fragile, provisional landing site where imported technologies pause, encounter local practices and are reshaped through cultural negotiation. Drawing on ethnographic fieldwork during the transfer of a Virtual Reality (VR) training tool from Airbus in France to CTRM in Malaysia's aerospace sector, the chapter explores how cultural logics—silence, tacit apprenticeship and group-based learning—clashed with the system's design assumptions of individual exploration, linear progression and direct feedback. The islet reframes such frictions not as failures but as resources for adaptation. Through vignettes of training rooms, observations of everyday negotiations and accounts of interface redesign, the chapter shows how engineers, managers and technicians collaboratively re-scripted the VR system to better align with Malaysian workshop culture. The *islet* is presented both as a conceptual and methodological tool. It illuminates the overlooked middle space of transfer where adaptation occurs, and it highlights the role of mediators who translate, negotiate and build trust between system and setting. The chapter argues that successful transfer depends less on plug-and-play universality than on cultivating adaptive landing zones. Technologies only take root when they learn the language of their users. The *Anthropotechnological Islet* is where that learning begins.

Keywords Anthropotechnology · Technology transfer · Cultural adaptation · Virtual reality · Tacit knowledge

Introduction

It was a humid afternoon when I first walked into the VR training room at AMIC's office in Bangi. The room looked as though it had been carefully staged: the polished headsets sat on a table draped in black cloth, cables neatly coiled and a large projector displayed the VR environment waiting to be explored. The Airbus engineers had

N. Ashikin Said, *Cultural Translation in Technology Design*,
SpringerBriefs in Anthropology, https://doi.org/10.1007/978-3-032-13751-7_9

already run several successful demos in Europe; there was a sense of quiet confidence in the air.

But when the first technician put on the headset, the mood shifted. He lasted perhaps two minutes before lifting the visor, chuckling to his colleagues: *"Pening la, macam sekolah je suara dia." ("It makes me dizzy, and the voice sounds like a schoolteacher").*Laughter rippled through the group. Another technician reached for the familiar paper manual, flipping through it while the headset lay idle on the table. Jokes about the "schoolteacher's voice" revealed that the neutral, formal tone of the interface was anything but neutral. This chapter grows out of that disjuncture. It explores the fragile in-between zone where technologies arrive but do not yet belong.

I call this zone the Anthropotechnological Islet: a temporary landing site where imported systems and local practices meet, hesitate and reshape one another.

The term is not accidental. Islands evoke stability and autonomy. An *islet*, by contrast, is smaller, less secure, provisional. It is a place one can land, but not necessarily settle. Transfer, I argue, does not unfold across smooth bridges or permanent islands; it depends on building these fragile islets; spaces of pause negotiation and adaptation—again and again, in every new context.

Conceptualizing the Anthropotechnological Islet

The idea of the Anthropotechnological Islet emerges at the intersection of several traditions: anthropotechnology, the sociology of technology and cultural cognition. The late French ergonomist Alain Wisner (1995) coined anthropotechnology to describe the design of technologies not for "humans in the abstract" but for specific cultural, social and cognitive realities. His case studies showed how machines that worked smoothly in Europe faltered elsewhere unless adapted to local conditions. For Wisner, the challenge was to move beyond universalist assumptions and to recognize cultural variability as a design parameter.

The islet is one way to operationalize Wisner's vision: it is the site where design and culture negotiate their encounter. Madeleine Akrich's (1992) concept of "scripts" helps explain what happens when technologies travel. Every system encodes assumptions about its users; their knowledge, rhythms, competencies and behaviors. The Airbus VR system carried a script of European workshop training: an individual learner, self-directed, navigating a linear progression of tasks with direct feedback. At CTRM, that script collided with a different one: collective learning, tacit apprenticeship, respect for hierarchy and silence as a form of communication. The islet is the stage where these scripts are confronted, contested and re-scripted.

Polanyi's (1966) notion of tacit knowledge reminds us that much of what workers know is not easily codified into manuals or modules. Nonaka and Takeuchi's (1995) knowledge creation model similarly emphasizes the interplay between tacit and explicit knowledge. In Malaysia's aerospace workshops, tacit apprenticeship through learning by observing seniors, mimicking gestures and absorbing work rhythms was central. The VR tool, by contrast, foregrounded explicit instruction. The islet becomes the place where tacit and explicit modes are translated into one another.

Everett Rogers (2003) argued that innovations spread not simply because they are superior but because they are compatible with existing practices, values and social structures. The plug-and-play fantasy of universal transfer ignores this. The islet, by foregrounding compatibility as an ongoing negotiation, reframes diffusion as co-creation. Research in cultural cognition has shown that cognitive performance is often context-dependent (Kozhevnikov, 2010). Our use of the PTA test confirmed this: technicians struggled with abstract spatial tasks but performed strongly when tasks were embedded in familiar workshop layouts. The islet, therefore, is also a cognitive scaffold anchoring abstract systems in embodied contexts.

During one session, a technician paused midway through the VR task, clearly unsure of the next step. Instead of asking the system, he turned to glance at a senior colleague standing nearby. The colleague raised his eyebrows slightly, a gesture of tacit approval and the technician continued. To an outside observer, this moment might look like inefficiency. But in that tiny glance, authority was checked, risk was shared and confidence was secured. This micro-practice was part of the islet: the fragile, culturally grounded scaffold that enabled technology to take root.

Mediator Roles in the Anthropotechnological Islet

The central theoretical contribution of this study lies in the development of an original concept—the anthropotechnological islet. This concept extends and refines earlier formulations of the anthropotechnological island and the anthropotechnological archipelago, offering a more nuanced lens for understanding transcultural technology transfer.

The anthropotechnological islet refers specifically to a small, cohesive group at the management and design level; in this case, the AMIC engineers—who provided the platform for the VR training tool and integrated the technology within the laboratory setting. Unlike the technicians and end-users, this group shared not only a high educational background but also international exposure, particularly to French culture. Many had studied in France, while others trained in the United Kingdom or the United States, and all were fluent in French. Although they were local Malaysians, their professional and cultural repertoires aligned closely with the expectations of Airbus.

This unique positioning rendered the management team a kind of "hybrid persona," simultaneously carrying French and Malay characteristics. Their familiarity with both worlds enabled them to act as cultural mediators: during meetings or casual interactions with Airbus representatives, they would converse in French, exchange jokes intelligible within the French cultural frame and ultimately create an atmosphere of familiarity. For Airbus, it was as if their "own people" were already present in Malaysia.

Within the framework of the anthropotechnological island, such an arrangement reflects the corporate ideal: a stable enclave where the technical and cultural ecosystem mirrors that of the technology provider. However, this stability proves fragile. When the VR training tool moved beyond the islet and reached the technicians (the ultimate users) the transfer was mediated by individuals with very different educational and cultural backgrounds. At this point, the "island" fragmented, giving way to an "archipelago": a constellation of smaller islands, each grounded in the French core technology but adapted to local Malaysian practices and cultural logics.

Figure 9.1 visualizes the theoretical movement from "islet" to "island" and ultimately to "archipelago." Each element represents a distinct scale and function in the dynamics of transcultural technology transfer:

1. **The anthropotechnological islet**: At the center lies the islet: the AMIC management engineers. These actors form a cohesive unit that is culturally hybrid—Malaysian by nationality yet trained, educated and socialized into French (and other Western) work cultures. The islet represents the initial stabilizing force of the transfer, as it mirrors the host's technical and cultural environment while simultaneously belonging to the recipient country.
2. **The anthropotechnological island**: Surrounding the islet is the broader construct of the island, which represents the corporate ideal of the provider

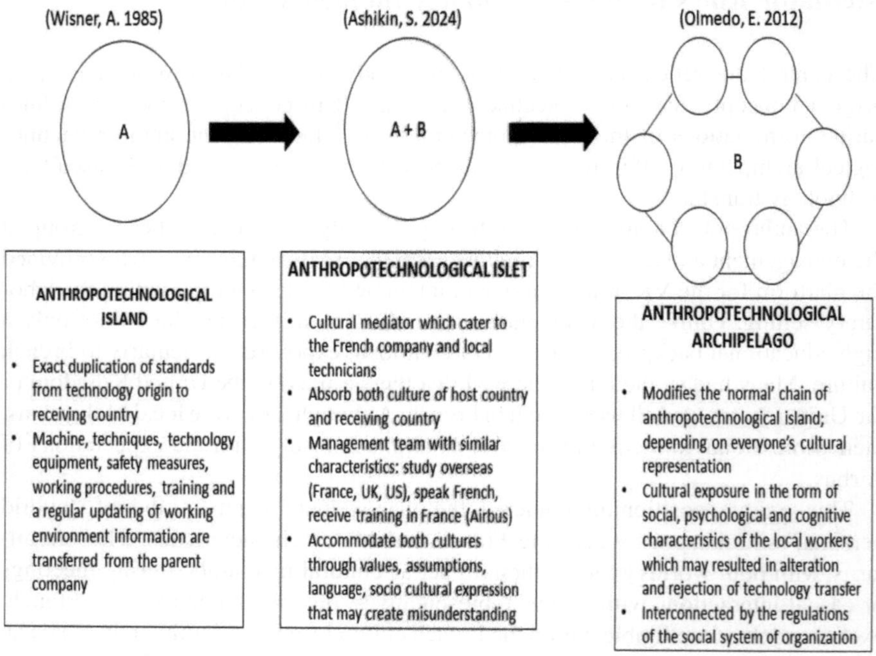

Fig. 9.1 The anthropotechnological islet in relation to the island and archipelago. (Author's own 2024)

(Airbus). Here, systems, norms and practices are reproduced almost intact, as though transplanted wholesale into a new environment. The island is stable as long as the transfer remains within the islet's circle, where the shared codes of language, training and professional ethos align closely with the host culture.

3. **The anthropotechnological archipelago**: Once the technology extends beyond the islet and the island—reaching technicians and shop-floor users—the situation changes. The single island fragments into an archipelago: multiple smaller sites where the French-origin technology is reinterpreted, adapted and negotiated according to local Malaysian contexts. Each island in the archipelago retains the technological "core" from Airbus but exhibits cultural inflections that reshape how the system is used, understood and sustained.

This schematic emphasizes that the "islet" is not the final destination but the crucial hinge: it ensures that the initial transfer is intelligible and credible to both sides, while also setting the conditions for subsequent diffusion into the more heterogeneous environment of the archipelago. Without the "islet," the "island" would collapse too quickly; without the "archipelago," the technology would never take root locally. Together, they form a continuum of adaptation and mediation that defines the social life of technology transfer.

The diffusion of technology thus scattered across multiple sites, shaped by non-technical factors such as language, education and workplace norms. The anthropotechnological islet is therefore central to the management of transcultural technology transfer. Positioned between the host (Airbus) and the recipient (CTRM technicians), the islet functions as a cultural mediator that absorbs, interprets and reconciles the expectations of both sides. Its mediating role can be delineated across several interrelated functions:

1. **Facilitating communication**: Acting as translators and interlocutors, mediators ensure that technical information, concerns and intentions are understood by both provider and recipient.
2. **Project management**: Beyond cultural tasks, mediators coordinate project timelines, milestones and deliverables, ensuring commitments are met.
3. **Understanding stakeholder needs**: They assess and balance the goals, constraints and expectations of both parties, tailoring the transfer process accordingly.
4. **Knowledge translation**: Mediators convert highly specialized technical knowledge into practical insights that make sense in the recipient's context.
5. **Building trust**: They establish credibility and foster confidence, creating a relational basis for successful transfer.
6. **Monitoring and evaluation**: Finally, they track whether the technology is used effectively, identifying adjustments where necessary.

In sum, the anthropotechnological islet encapsulates the paradox of technology transfer: it is simultaneously a space of stability, where shared cultural codes ease the transfer and a fragile node, vulnerable to fragmentation once knowledge circulates beyond its immediate boundaries. Its existence demonstrates how technology transfer is never solely technical but fundamentally cultural, requiring mediation, negotiation and adaptation across scales.

If the "islet" is the landing place, mediators are its builders. Technologies do not adapt themselves; nor do cultural practices automatically yield. It is people (engineers, managers, trainers, even technicians) who translate, negotiate and sustain the fragile middle. The literature on boundary objects (Star & Griesemer, 1989), cultural brokerage (Wolf, 1956; Wenger, 1998) and technology mediation (Oudshoorn & Pinch, 2003) reminds us that technologies travel only through such human work.

In our project, I identified seven core mediator roles. While analytically distinct, in practice they often overlap. Each role is illustrated here with a vignette from the VR training experience.

1. **Translation**: Translation is the first and most obvious role. It is not simply about switching words from French to Malay or English. It is about translating logics, metaphors and forms of authority. Latour (1987) reminds us that translation is always transformation: messages change as they move across contexts.
2. **Adaptation**: Adaptation involves tweaking not just language but the structure of interaction. Anthropologists of work (Suchman 1987; Orr 1996) have shown that people rarely follow systems as prescribed; they adapt them in practice. Mediators formalize this adaptation, adjusting the system to fit local rhythms. In early trials, the VR module required trainees to complete tasks alone. But in CTRM, silence and group-based observation were core to learning. Trainers improvised: instead of sending technicians one by one, they allowed small groups to watch together, commenting quietly. Later, engineers re-scripted the software to include pause points where discussion was encouraged. Adaptation, in this sense, was not a bug but a feature. It showed how mediators made the VR tool bend toward tacit apprenticeship without losing its instructional rigor.
3. **Trust:** Trust is the lubricant of transfer. New technologies often generate suspicion. Are they replacing jobs, exposing incompetence or enforcing surveillance? Fukuyama (1995) highlights trust as a foundation of organizational cooperation. Mediators build trust by humanizing the system, ensuring that it is not perceived as a threat. One technician confided quietly after a session: *"Kalau VR ni rekod semua salah saya, nanti bos nampak ke?" ("If this VR records all my mistakes, will my boss see them?")* He was worried the tool might expose weaknesses to management. A mediator—an engineer who had studied in France but spoke fluent Malay—reassured him: "*Tak, ni untuk belajar je, bukan untuk markah.*" ("*No, this is only for learning, not grading.*") That reassurance transformed participation. The following week, the same technician volunteered to try again, this time staying in the headset longer. Trust, carefully built, allowed the islet to stabilize.
4. **Negotiation:** Negotiation is the active balancing of interests: between efficiency and comfort, standardization and flexibility, provider and recipient. Negotiation aligns with Callon's (1986) idea of "translation" as enrollment, convincing actors to align with a shared project. Mediators become negotiators who broker compromises. CTRM trainers requested that the narration remain in Malay to

match everyday practice. After several meetings, a compromise was reached: they use a specific communication code that includes English and specific terms (mixture of English and Malay) that the shop-floor technicians use on everyday basis. This negotiation did not resolve the tension once and for all; it institutionalized flexibility. The islet, in this case, was co-constructed through ongoing bargaining.

5. **Boundary Spanning:** Boundary spanners (Aldrich & Herker, 1977; Levina & Vaast, 2005) operate across organizational and cultural borders. They are "insiders" in more than one world: familiar with Airbus's technical logics yet at ease in Malaysian workshop culture. Their hybrid identities allow them to ferry meanings across divides. One AMIC engineer exemplified this role. He studied for several years in France, spoke fluent French and shared jokes with visiting Airbus staff. But meeting the local technicians, he switched seamlessly into colloquial Malay, chatting about family and football with technicians. In meetings, he interpreted not just words but intentions: when Airbus managers pushed for "faster adoption," he reframed it as "building confidence step by step" for local trainers. His dual credibility made him indispensable. Without such boundary spanners, the islet would have collapsed under cultural misalignment.
6. **Advocacy:** Mediators often act as advocates for the voiceless. Technicians at the bottom of the hierarchy may not feel comfortable raising concerns. Advocacy ensures that their lived realities inform system redesign. In actor-network terms, mediators "speak for" absent actors (Callon, 1986).
7. **Feedback Loops:** Finally, mediators create feedback loops: structured channels where user experiences travel back to designers and design adjustments return to users. Without loops, transfer stagnates. Feedback embodies Nonaka's (1995) spiral of knowledge creation; moving between tacit and explicit. We instituted casual review sessions where technicians, trainers and engineers gathered informally over tea. Instead of formal reports, mediators encouraged technicians to share anecdotes when the headset slipped, when dizziness set in or when laughter erupted. Engineers took notes and within weeks subtle tweaks appeared: lighter straps, shorter modules, new prompts. This rhythm of feedback created a living islet: never finished, always adapting.

The seven roles (translation, adaptation, trust, negotiation, boundary spanning, advocacy and feedback loops) show that mediation is not peripheral to transfer. It is the transfer. The islet is not a passive space; it is built and rebuilt through these roles. One evening, after a long day of training, I overheard two technicians discussing the VR tool. One said, half-jokingly: "*Mula-mula macam benda orang putih. Sekarang rasa macam benda kita juga.*" ("*At first it felt like a foreign gadget. Now it feels like ours too*") That small shift from foreign to ours was the islet in action.

When I first began thinking of the Anthropotechnological Islet, I imagined it as a metaphorical landing dock: a place where technologies arrive but cannot move further without careful handling. Over time, I realized that the dock was not just a passive space, but it was populated by people acting in distinctive roles, often without being aware that they were the crucial link keeping the technology afloat.

Taken together, these seven roles form the invisible infrastructure of the Anthropotechnological Islet. They are not always performed by the same person, nor are they formally assigned. Instead, they emerge situationally, often improvisationally, in response to gaps, frictions and silences.

What struck me most was that the islet was never static. It was constantly shifting, depending on who was present, what was at stake and how much risk people were willing to bear in making themselves understood. In this sense, the islet was less like a dock carved in stone and more like a raft built and rebuilt from whatever materials were at hand—translation here, trust there, a plank of advocacy, a rope of negotiation.

Global Resonance: Beyond Malaysia

While my fieldwork was rooted in the hangars of Melaka and the training rooms of AMIC, it soon became clear that the dynamics I witnessed were not unique to Malaysia. Technology transfer has always been a global drama, played out on different stages with varying casts. Yet the themes of mediation, silence and improvisation recur with striking familiarity.

The islet, once again, appeared in a different guise: a space where mediators enabled crossing, but where their presence remained invisible in official reports. Anthropotechnological Islet is not confined to one country, one industry, or one era. It is a recurring condition of transfer wherever technology crosses borders. What differs are the materials from which the islet is built: in Brazil, bilingual engineers; in Kenya, community interpreters; in Indonesia, pedagogical metaphors; in Malaysia, a whispered explanation and a cigarette-break confession. If the islet is everywhere, why does it so often go unrecognized? One answer lies in the dominant imaginaries of technology transfer, which emphasize hardware, investment and intellectual property, while sidelining the human work of mediation.

Policymakers often treat mediators as auxiliary rather than essential, resulting in chronic underinvestment in training, trust-building and cultural brokerage. In Malaysia, I often heard officials speak of "absorptive capacity" as if it were a measurable attribute of organizations, akin to financial capital or machine hours. But in practice, absorptive capacity depended on whether a technician felt safe enough to ask a question, or whether a supervisor cared enough to rephrase a point for clarity. These micro-level acts are hard to quantify, yet they are the very substance of transfer.

There are also ethical stakes. Whose voice is amplified in feedback loops, and whose is silenced? Who has the authority to negotiate adaptations and who must simply endure discomfort? The vignette of the hijab-wearing engineer illustrates this: without advocacy, her embodied experience would have remained invisible, subsumed under the rhetoric of "universal ergonomics." To ignore such voices is not only inefficient but unjust. Recognizing the islet means acknowledging that technology transfer is never value-neutral. It is always political, shaped by asymmetries of power, gender and language. For anthropologists of technology, this recognition demands a methodological humility: to pay attention not only to machines and protocols but to the whispered translations, the improvised fans, the cigarettes shared in silence.

Conclusion

When I first coined the metaphor of the Anthropotechnological Islet, I imagined it as a fixed dock, a place where technologies arrive and wait for clearance. After months of fieldwork, I realized it was more fluid than that—less a dock and more a raft. It floated, shifted and was rebuilt constantly from the materials at hand; a word of translation, a gesture of trust, a quiet act of advocacy.

The Airbus–CTRM partnership did not succeed or fail because of the intrinsic quality of the VR headset. It survived because mediators that are often invisible, often undervalued, made the crossing possible. They spanned boundaries, negotiated compromises, built trust and fed back insights. Without them, the technology would have remained stranded, like a ship anchored offshore with no means of landing. To speak of an Anthropotechnological Islet is, therefore, to insist that technology transfer is not only about machines but about people. It is about the fragile, improvised, yet enduring structures they build to allow technology to take root in new soil.

And if we want transfer to work—not just in aerospace, not just in Malaysia, but anywhere—we must learn to see, support and honor the mediators who make the crossing possible. The "islet," in the end, is less a metaphor than a reality: a space of encounter, tension and passage, where the human and the technical meet. And like all "islets," it is small, precarious and easily overlooked. But without it, there is no landing.

References

Akrich, M. (1992). The de-scription of technical objects. In W. E. Bijker & J. Law (Eds.), *Shaping technology/building society: Studies in sociotechnical change* (pp. 205–224). MIT Press.

Callon, M. (1986). Some elements of a sociology of translation: Domestication of the scallops and the fishermen of St Brieuc Bay. In J. Law (Ed.), *Power, action and belief: A new sociology of knowledge?* (pp. 196–233). Routledge.

Kozhevnikov, M. (2010). Cognitive styles in the context of modern psychology: Toward an integrated framework of cognitive style. *Psychological Bulletin, 136*(3), 464–481.

Levina, N., & Vaast, E. (2005). The emergence of boundary spanning competence in practice: Implications for implementation and use of information systems. *MIS Quarterly, 29*(2), 335–363.

Nonaka, I., & Takeuchi, H. (1995). *The knowledge-creating company: How Japanese companies create the dynamics of innovation.* Oxford University Press.

Oudshoorn, N., & Pinch, T. (Eds.). (2003). *How users matter: The co-construction of users and technologies.* Cambridge, MA: MIT Press.

Star, S. L., & Griesemer, J. R. (1989). Institutional ecology, 'translations' and boundary objects: Amateurs and professionals in Berkeley's Museum of Vertebrate Zoology, 1907–39. *Social Studies of Science, 19*(3), 387–420.

Wenger, E. (1998). *Communities of practice: Learning, meaning, and identity*. Cambridge: Cambridge University Press.

Wolf, E. R. (1956). Aspects of group relations in a complex society: Mexico. *American Anthropologist, 58*(6), 1065–1078.

Chapter 10
Conclusion

Abstract This concluding chapter synthesizes the book's central insights on technology transfer, culture and innovation through the lens of the Malaysian aerospace case. Beginning with the puzzle of why a sophisticated Airbus VR training tool faltered at CTRM despite its technical robustness, the chapter argues that transfer is never purely technical but always cultural and epistemic. The lesson is that technology does not travel as neutral hardware. It carries cultural scripts that can either enable or derail transfer. To capture these dynamics, the chapter introduces the concept of the anthropotechnological islet: fragile, generative micro-worlds of managers, engineers and designers who mediate between global artifacts and local practices. Building on Eric Olmedo's metaphor of the anthropotechnological archipelago, the islet refines the analytical focus to the meso level, illuminating the often invisible work of translation, adaptation and cultural brokerage. Methodologically, the chapter reflects on ethnography's unique role in exposing tacit knowledge and interactional expertise that quantitative or policy-driven approaches overlook. In terms of policy and practice, the islet highlights the need to support micro-sites of negotiation rather than assume that agreements and infrastructures alone guarantee success. Looking forward, the chapter situates the islet within wider debates about digitalization, artificial intelligence and global South industrialization, arguing that the future of technology transfer depends on cultivating fragile but indispensable zones where humans and machines co-produce new worlds.

Keywords Anthropotechnological islet · Technology transfer · Ethnography of expertise · Global South industrialisation · Human–machine collaboration

Introduction

This book has taken the reader on a journey across shifting terrains of technology transfer, anthropology of expertise and the fragile negotiations of culture in a Malaysian aerospace case. Beginning with the empirical puzzle of why a sophisticated Airbus VR training tool faltered in its transfer to CTRM, the study has

N. Ashikin Said, *Cultural Translation in Technology Design*,
SpringerBriefs in Anthropology, https://doi.org/10.1007/978-3-032-13751-7_10

unfolded into a theoretical reflection on the entanglement of humans, technologies and institutions in processes of global industrial exchange.

This chapter closes the circle. It returns to the original problem—what makes technology transfer succeed or fail—but does so from a new vantage point. By now the reader has seen that transfer is neither purely technical nor purely social. It is instead an ongoing negotiation, mediated by what I have conceptualized as the anthropotechnological islet: a micro-configuration of managers, engineers and designers whose interactions generate a platform for innovation, adaptation and sometimes, breakdown. The islet emerges within the broader sea of global capitalism, nestled between larger structures of power and institutional archipelagos. It is both fragile and generative, contingent yet indispensable.

This final chapter synthesizes the empirical insights and theoretical interventions of the book. It does so in four movements. First, revisiting the empirical contribution of the study, showing how the CTRM–Airbus case unsettles dominant assumptions in technology transfer research. Second, zooming in on the theoretical significance of the anthropotechnological islet as an original concept that extends the anthropology of technology. Third, I reflect on the methodological implications for doing ethnography at the intersection of global industry and local expertise. Finally, I look forward: situating the lessons of this study within wider debates about global South industrialization, digitalization and the uncertain futures of human–machine collaboration.

Reframing Technology Transfer: Beyond Success and Failure

Technology transfer has long been analyzed through the lens of capacity building, absorptive ability and institutional readiness (Bell & Pavitt, 1993; Lall, 1992). Much of the literature presumes a linear process: a technology is designed in one place, exported to another and then either adopted successfully or rejected. Metrics of success often focus on productivity gains, skill acquisition or competitiveness in global value chains (Dahlman et al., 1987). While valuable, such frameworks obscure the everyday frictions that occur at the point of encounter between technologies and their users.

The Airbus VR training tool at CTRM is a case in point. Technically, the system was flawless. It arrived without malfunction, having been validated in European aerospace environments. Yet on the shop floor in Melaka, it faltered. Not because of software bugs or hardware defects, but because its design carried tacit cultural assumptions about how technicians should learn, how their spatial cognition was structured and how authority was exercised in training regimes. The Malaysian technicians who donned the headset did not lack intelligence or willingness. They lacked alignment with a training device encoded with foreign pedagogical logics.

This empirical detail unsettles two common presumptions.

First, that technology transfer is primarily a matter of skill deficits to be remedied by training. Second, that culture is an external barrier to be managed rather than an intrinsic dimension of technological design. The CTRM–Airbus case shows that transfer is not merely the movement of artifacts but the movement of ways of knowing, ways of sensing and ways of ordering work. In short, transfer is epistemic and cultural before it is technical.

By tracing this dynamic, the book contributes to a growing scholarship that emphasizes the sociocultural embedding of technology (Orr, 1996; Suchman, 2007; Vertesi, 2014). In particular, it supports the argument that technologies are not neutral carriers of function but vehicles of culture (scripts, values and assumptions) inscribed into hardware and software alike (Akrich, 1992). The Airbus VR tool bore the imprint of its European engineering environment, and its misfit at CTRM illustrates how transfer failures often originate in design choices invisible to their creators.

The Anthropotechnological Islet: Conceptualizing Micro-worlds of Transfer

The key theoretical contribution of this book is the concept of the anthropotechnological islet. Building on but also revising earlier metaphors of islands and archipelagos, the islet offers a lens to see how technology transfer actually occurs in practice. Previous scholarship has employed the metaphor of the archipelago to describe the dispersed yet connected sites of contemporary globalization (Helmreich, 2009; Ong & Collier, 2005; Rabinow, 2003). Within this lineage, Olmedo (2015) advanced the notion of the anthropotechnological archipelago, a conceptual framework for understanding how humans and technologies co-exist across scattered but interlinked domains.

He uses the metaphor of the workplace as a micro-ecumene, showing how identity, power and spatial practices converge in everyday training and the transfer of workplace norms. His work provides a point of departure for the concept of technological archipelagos and by extension, the refinement of those archipelagos into what I term the "anthropotechnological islet." Olmedo's archipelago foregrounds the hybridization of technical and social practices, and the necessity of navigating cultural frontiers in technological encounters. This imagery captures the sense that expertise, capital and knowledge flow between nodes of varying scale.

Yet the metaphor often remains at the macro-level, focusing on national innovation systems or transnational networks. Building on this foundation, the anthropotechnological islet zooms in to a smaller scale. If the archipelago maps a constellation of dispersed sites, the islet identifies the micro-worlds within those sites—the fragile clusters of managers, engineers and designers who mediate transfer in practice. In the case of CTRM, the islet was constituted by the management team tasked with implementing the VR tool and the Airbus engineers who designed it. They were the ones who negotiated adaptation, translated instructions and patched over misalignments.

Without their micro-world of mediation, the transfer would have either collapsed or stagnated. There are three features that define the islet:

1. **Proximity**: Members of the islet are close enough, organizationally or socially, to directly influence the transfer process. They are not distant policymakers or end-users but immediate negotiators.
2. **Hybrid Expertise**: Islet members combine technical knowledge with cultural or organizational navigation. The AMIC engineers, for instance, not only understood the VR tool's mechanics but also grasped how CTRM technicians approached learning.
3. **Fragility**: The islet is not institutionalized. It can dissolve with personnel changes, managerial reshuffling or shifting corporate priorities. Its generative power is contingent.

The anthropotechnological islet extends the anthropology of technology by offering a meso-level analytical unit. It is neither the grand structure of global capitalism nor the intimate detail of individual practice. It is a micro-world that makes transfer possible, yet is often invisible in both managerial reports and ethnographic accounts. By naming it, this book provides scholars and practitioners with a new vocabulary to apprehend where and how transfer actually takes shape. The islet is a lens to see the otherwise unseen labor of mediation, translation and adaptation that underpins the circulation of technology in the global South and beyond.

Methodological Reflections: Ethnography at the Edge of Industry

This book has also sought to demonstrate what an ethnographic approach can contribute to technology transfer studies. Much of the existing literature is dominated by economics, management and policy analysis. These approaches yield important insights into absorptive capacity, firm strategy and institutional environments. Yet they often miss the texture of lived experience and the subtle negotiations of meaning that occur on the shop floor.

By embedding ethnography within CTRM's environment, I was able to capture technicians' frustrations with the VR headset, managers' anxieties about performance indicators and engineers' pride in their design work. These are not anecdotal embellishments but central to explaining why the transfer faltered. They reveal that transfer is not only about systems alignment but about human sense-making. Methodologically, this raises questions about the positionality of the researcher. As an anthropologist working in a high-tech industrial setting, I was both insider and outsider: granted access to meetings, yet never fully part of corporate decision-making. This liminal position mirrors the islet itself—a small space of mediation between worlds.

Indeed, the research process can be understood as a kind of islet: a fragile zone where knowledge is translated between anthropology and engineering, between global theories and local practices. For future research, this suggests that ethnography has a unique role in revealing the tacit dimensions of transfer—what Polanyi (1966) called "tacit knowledge," and what Collins (2010) later described as the "interactional expertise" necessary for knowledge circulation. Without ethnography, such dimensions remain invisible. With it, we gain a fuller picture of what makes transfer succeed or fail.

Implications for Malaysia and the Global South

The CTRM–Airbus case is not only about one company or one tool. It is emblematic of broader dynamics faced by industrializing economies in the global South. Malaysia has long sought to move up the value chain, from assembly to design, from low-wage labor to high-skill innovation (Rasiah, 2010). Technology transfer is central to this ambition. Yet as the VR case shows, the challenge is not only infrastructural or educational but cultural and epistemic.

If transfer is mediated by anthropotechnological islets, then policy must recognize and support these fragile sites. This means investing not only in hardware and training programs but in cultural brokerage, interdisciplinary teams and organizational spaces where engineers, managers and technicians can negotiate together. Too often, industrial policy assumes that if infrastructure is built and agreements are signed, transfer will follow. The islet metaphor warns us otherwise: transfer depends on the cultivation of micro-worlds where human–technology relations are actively aligned.

Looking Forward: Anthropotechnologies in the Digital Age

As the world moves deeper into the era of digitalization, artificial intelligence and automation, the questions raised by this book acquire new urgency. If a VR headset already embeds cultural assumptions that can disrupt transfer, what of machine learning systems, predictive algorithms or fully automated production lines? These are not merely tools but epistemic infrastructures—devices that encode not only ways of learning but ways of seeing, deciding and governing.

The anthropotechnological islet becomes even more crucial in this context. For as technologies grow more complex and opaque, the need for micro-worlds of mediation intensifies. Someone must translate between algorithmic logics and human norms, between global design templates and local practices. Without such islets, digitalization risks reproducing forms of dependency and exclusion, where the global South becomes not a partner in innovation but a passive user of foreign epistemologies. This is not a counsel of despair but of vigilance. The islet concept provides a hopeful lens: it shows that even in asymmetrical conditions, small groups of actors can carve out spaces of negotiation, adaptation and creativity. They can re-script technologies, re-align practices and even innovate in return. The future of technology transfer, then, lies not in erasing cultural difference but in cultivating these fragile zones of encounter.

The Islet as Metaphor, Method and Ethic

As we draw this book to a close, let me return to the islet. It is more than a metaphor. It is also a method and an ethic.

1. As a metaphor, it helps us see the micro-worlds where transfer happens, offering a vocabulary for what is otherwise overlooked.
2. As a method, it guides researchers to zoom in to study not only macro-flows of capital, but the small sites of mediation that make such flows possible.
3. As an ethic, it reminds us of fragility. Islets are vulnerable to neglect, dissolution and capture. To support technology transfer responsibly means to nurture these fragile zones, to protect the human labor of mediation that sustains them.

The Airbus–CTRM case has shown that when islets are overlooked, transfer falters. When they are cultivated, transfer can flourish—not as a perfect linear success, but as a negotiated co-production. In this sense, the islet is not only an analytical tool but a call to practice: a reminder that the future of global technology transfer depends on how we care for the small worlds where humans and machines meet.

References

Akrich, M. (1992). The description of technical objects. In W. Bijker & J. Law (Eds.), *Shaping technology/building society: Studies in sociotechnical change* (pp. 205–224). MIT Press.

Bell, M., & Pavitt, K. (1993). Technological accumulation and industrial growth: Contrasts between developed and developing countries. *Industrial and Corporate Change, 2*(2), 157–210.

Collins, H. (2010). *Tacit and explicit knowledge*. University of Chicago Press.

Dahlman, C. J., Ross-Larson, B., & Westphal, L. E. (1987). Managing technological development: Lessons from the newly industrializing countries. *World Development, 15*(6), 759–775.

Helmreich, S. (2009). *Alien ocean: Anthropological voyages in microbial seas*. University of California Press.

Lall, S. (1992). Technological capabilities and industrialization. *World Development, 20*(2), 165–186.

Olmedo, E. (2015). *Identity at work: Ethnicity, food & power in Malaysian hospitality industry*. Springer Singapore. https://doi.org/10.1007/978-981-287-561-7

Ong, A., & Collier, S. J. (2005). *Global assemblages: Technology, politics, and ethics as anthropological problems*. Blackwell.

Orr, J. (1996). *Talking about machines: An ethnography of a modern job*. Cornell University Press.

Polanyi, M. (1966). *The tacit dimension*. Routledge.

Rabinow, P. (2003). *Anthropos today: Reflections on modern equipment*. Princeton University Press.

Rasiah, R. (2010). Are electronics firms in Malaysia catching up in the technology ladder? *Journal of the Asia Pacific Economy, 15*(3), 301–319.

Suchman, L. (2007). *Human–machine reconfigurations: Plans and situated actions* (2nd ed.). Cambridge University Press.

Vertesi, J. (2014). *Seeing like a rover: How robots, teams, and images craft knowledge of Mars*. University of Chicago Press.

Index

N. Ashikin Said, *Cultural Translation in Technology Design*, SpringerBriefs in Anthropology, https://doi.org/10.1007/978-3-032-13751-7

MIX
Papier aus verantwortungsvollen Quellen
Paper from responsible sources
FSC® C105338

If you have any concerns about our products,
you can contact us on
ProductSafety@springernature.com

In case Publisher is established outside the EU,
the EU authorized representative is:
Springer Nature Customer Service Center GmbH
Europaplatz 3, 69115 Heidelberg, Germany

Printed by Libri Plureos GmbH
in Hamburg, Germany